SOLAR ENERGY

APPLICATION, ECONOMICS, AND PUBLIC PERCEPTION

SOLAR ENERGY

APPLICATION, ECONOMICS, AND PUBLIC PERCEPTION

Edited by
Muyiwa Adaramola, PhD

Apple Academic Press

TORONTO   NEW JERSEY

CRC Press
Taylor & Francis Group
6000 Broken Sound Parkway NW, Suite 300
Boca Raton, FL 33487-2742

Apple Academic Press, Inc
3333 Mistwell Crescent
Oakville, ON L6L 0A2
Canada

© 2015 by Apple Academic Press, Inc.

First issued in paperback 2021

Exclusive worldwide distribution by CRC Press an imprint of Taylor & Francis Group, an Informa business

No claim to original U.S. Government works

Version Date: 20141010

ISBN 13: 978-1-77463-233-8 (pbk)
ISBN 13: 978-1-77188-090-9 (hbk)

Visit the Taylor & Francis Web site at
http://www.taylorandfrancis.com

and the CRC Press Web site at
http://www.crcpress.com

For information about Apple Academic Press product
http://www.appleacademicpress.com

ABOUT THE EDITOR

MUYIWA ADARAMOLA, PhD

Dr. Muyiwa S. Adaramola earned his BSc and MSc in Mechanical Engineering from Obafemi Awolowo University, Nigeria, and University of Ibadan, Nigeria, respectively. He received his PhD in Environmental Engineering at the University of Saskatchewan in Saskatoon, Canada. He has worked as lecturer at the Obafemi Awolowo University and as a researcher at the Norwegian University of Science and Technology, Trondheim, Norway. Currently, Dr. Adaramola is an Associate Professor in Renewable Energy at the Norwegian University of Life Sciences, Ås, Norway.

CONTENTS

PART V: PUBLIC PERCEPTIONS OF SOLAR ENERGY

ACKNOWLEDGMENT AND HOW TO CITE

The editor and publisher thank each of the authors who contributed to this book, whether by granting their permission individually or by releasing their research as open source articles or under a license that permits free use, provided that attribution is made. The chapters in this book were previously published in various places in various formats. To cite the work contained in this book and to view the individual permissions, please refer to the citation at the beginning of each chapter. Each chapter was read individually and carefully selected by the editor; the result is a book that examines the use of solar energy from a variety of different perspectives, including technology, economics, and public perceptions.

LIST OF CONTRIBUTORS

Muyiwa S. Adaramola
Department of Ecology and Natural Resource Management, Faculty of Environmental Science and Technology, Norwegian University of Life Sciences, Ås, Norway

Aldo Arnone
The College, The University of Chicago, 5801 South Ellis Ave., Chicago, IL, USA

Vanessa Arellano Banoni
The College, The University of Chicago, 5801 South Ellis Ave., Chicago, IL, USA

D. Richard Cameron
The Nature Conservancy, San Francisco, California, United States of America

T. T. Chow
Building Energy and Environmental Technology Research Unit, Division of Building Science and Technology, City University of Hong Kong, Tat Chee Avenue, Hong Kong

Brian S. Cohen
The Nature Conservancy, San Francisco, California, United States of America

Paul Denholm
National Renewable Energy Laboratory, Colorado, United States of America

Yi Ding
Department of Electrical Engineering, Technical University of Denmark, Lyngby, Denmark

Simon Eckermann
Australian Health Services Research Institute, Sydney Business School, University of Wollongong, New South Wales, 2522, Australia

B.C. Farhar
University of Colorado at Boulder, Colorado, United States of America

Maria Fondeur
The College, The University of Chicago, 5801 South Ellis Ave., Chicago, IL, USA

Lalit Goel
School of EEE, Nanyang Technological University, Singapore, Singapore

Greg Hampton
Academic Services Division, University of Wollongong, New South Wales, 2522, Australia

Annabel Hodge
The College, The University of Chicago, 5801 South Ellis Ave., Chicago, IL, USA

Marissa Hummon
National Renewable Energy Laboratory, Colorado, United States of America

L.M. Hunter
University of Colorado at Boulder, Colorado, United States of America

T.M. Kirkland
University of Colorado at Boulder, Colorado, United States of America

Colin Klinger
Department of Physics and Astronomy, California State University Northridge, Northridge, California, United States of America

Lado Kurdgelashvili
Center for Energy and Environmental Policy, University of Delaware, 278 Graham Hall, Newark, DE 19716, USA

Gregory Levitin
Reliability & Equipment Department, R&D Division, The Israel Electric Corporation Ltd., Haifa, Israel

Mark Mehos
National Renewable Energy Laboratory, Colorado, United States of America

C. Menezo
CETHIL UMR 5008, Domaine Scientifique de La Doua, INSA de Lyon, 9 Rue de la Physique, 69621 Villeurbanne Cedex, France

Scott A. Morrison
The Nature Conservancy, San Francisco, California, United States of America

Patrick A. Narbel
Department of Finance and Management Science, Norwegian School of Economics and Business Administration, NHH, Helleveien 30, NO-5045 Bergen, Norway

J Patrick Offner
The College, The University of Chicago, 5801 South Ellis Ave., Chicago, IL, USA

Yogeshwari Patel
Department of Physics and Astronomy, California State University Northridge, Northridge, California, United States of America

Jordan K Phillips
The College, The University of Chicago, 5801 South Ellis Ave., Chicago, IL, USA

Henk W. Ch. Postma
Department of Physics and Astronomy, California State University Northridge, Northridge, California, United States of America

Weixiang Shen
Faculty of Engineering & Industrial Sciences, Swinburne University of Technology, Hawthorn, VIC, Australia

K.J. Tierney
University of Colorado at Boulder, Colorado, United States of America

Govinda R. Timilsina
Environmental and Energy Unit, Development Research Group, The World Bank, 1818 H Street NW, Washington, DC, USA

G. N. Tiwari
Centre for Energy Studies, Indian Institute of Technology Delhi, Hauz Khas, New Delhi 11 00 16, India

Peng Wang
School of EEE, Nanyang Technological University, Singapore, Singapore

Qiuwei Wu
Department of Electrical Engineering, Technical University of Denmark, Lyngby, Denmark

Shenyi Wu
Department of Architecture and Built Environment, University of Nottingham, University Park, Nottingham NG7 2RD, UK

Chenguang Xiong
Department of Architecture and Built Environment, University of Nottingham, University Park, Nottingham NG7 2RD, UK

INTRODUCTION

Due to global decline in fossil fuel reserves and contribution of their emissions (during the extraction, production and utilization processes) to climate change, many countries are now examining their national energy policies with a view of shifting toward low-carbon and renewable sources of energy. In addition, security of supply and fluctuations in crude-oil prices (which can be sensitive to internal and regional conflicts) can lead to economic vulnerability of oil-importing countries. Furthermore, the exporting countries can use the price of these resources to settle political differences.

As a result of recent trends in solar energy development due to improved technology, cost reduction in solar energy application equipment, and possible use with energy storage systems, solar energy is expected to play a significant role in the future global energy mix, both in the developed and developing countries. This book discusses the applications, economics, and public perceptions of solar energy conversion systems. The book is divided into five sections: the first section provides an overview of hybrid solar energy systems, the second discusses solar energy and conservation issues, the third focuses on solar energy technology, the fourth section is on the economics of solar energy, and, finally, the final section addresses public perceptions of solar energy.

In Chapter 1, Chow and colleagues give a broad review of the published academic works on hybrid photovoltaic/thermal (PVT) collector systems, with an emphasis placed on the research and development activities in the last decade.

In Chapter 2, Cameronand colleagues examine the synergy between renewable energy generation goals and those for biodiversity conservation in the Mojave Desert of the southwestern USA. They integrate spatial data on biodiversity conservation value, solar energy potential, and land surface slope angle (a key determinant of development feasibility) and found there to be sufficient area to meet renewable energy goals without developing on lands of relatively high conservation value. Indeed, they found nearly 200,000 ha of lower conservation value land below the most restrictive slope angle (<1%); that area could meet the state of California's current 33% renewable energy goal 1.8 times over. They found over 740,000 ha below the highest slope angle (<5%)—an area that can meet California's renewable energy goal seven times over. Their analysis also suggests that the supply of high quality habitat on private land may be insufficient to mitigate impacts from future solar projects, so enhancing public land management may need to be considered among the options to offset such impacts. Using the approach presented here, planners could reduce development impacts on areas of higher conservation value, and so reduce trade-offs between converting to a green energy economy and conserving biodiversity.

Klinger and colleagues presents proof-of-concept all-carbon solar cells in Chapter 3. These solar cells are made of a photoactive side of predominantly semiconducting nanotubes for photo-conversion and a counter electrode made of a natural mixture of carbon nanotubes or graphite, connected by a liquid electrolyte through a redox reaction. The cells do not require rare source materials such as In or Pt, nor high-grade semiconductor processing equipment. They do not rely on dye for photo-conversion and therefore do not bleach, and are easy to fabricate using a spray-paint technique. They observed that cells with a lower concentration of carbon nanotubes on the active semiconducting electrode perform better than cells with a higher concentration of nanotubes. This effect is contrary to the expectation that a larger number of nanotubes would lead to more photo-conversion and therefore more power generation. The authors attribute this to the presence of metallic nanotubes that provide short for photo-excited electrons, bypassing the load. They demonstrate optimization strategies that improve cell efficiency by orders of magnitude and conclude that, once it is possible to make semiconducting-only carbon nanotube films that may provide the greatest efficiency improvement.

In Chapter 4, Denholm and Mehos examines the degree to which concentrating solar power (CSP) may be complementary to PV via its use of thermal energy storage. The authors first review the challenges of PV deployment at scale with a focus on the supply/demand coincidence and limits of grid flexibility. They then perform a series of grid simulations to indicate the general potential of CSP with thermal energy storage (TES) to enable greater use of solar generation, including additional PV. The use of thermal energy storage in concentrating solar power plants provides one option for increased grid flexibility in two primary ways. First, TES allows shifting of the solar resource to periods of reduced solar output with relatively high efficiency. Second is the inherent flexibility of CSP/ TES plants, which offer higher ramp rates and ranges than large thermal plants currently used to meet a large fraction of electric demand. Finally, they use these reduced form simulations to identify the data and modeling needed for more comprehensive analysis of the potential of CSP with TES to provide additional flexibility to the grid as a whole and benefit all variable generation sources.

Adaramola examines the feasibility of solar PV-grid tied energy system for electricity generation in a selected location in the northern part of Nigeria in Chapter 5. The technical and economic performance of a combination of 80 kW solar PV-grid connected was investigated. The effects of the cost of PV system and global solar radiation were also investigated. Based on the findings reported in this chapter, the author concludes that the development of grid-connected solar PV system in the north-eastern part of Nigeria could be economically viable.

In Chapter 6, different cooling methods to reduce high ambient temperature effect on photovoltaic panels' performance were proposed and investigated. This chapter reviews the previous work on cooling PV cells and concludes that the cost-effectiveness, design feasibility and minimal energy consumption are the important design consideration for cooling systems. Based on these considerations, the authors report a passive cooling method that utilizes rainwater as cooling media and a gas expansion device to distribute the rainwater. The gas is thermally expanded from receiving solar radiation as such the amount of water it pushes to flow over the PV cells is proportional with the solar radiation it received. The chapter reports a design and simulation of such a system for a domestic house application and a relationship of the gas chamber size, solar radiation and gas expansion volume was established for evaluation with respect to the variation of gas temperature and the amount of rainwater used for cooling. A heat transfer model was used to evaluate the performance of the cells by cooling with this passive device. The results show that on a design day, the passive cooling system reduces the temperature of the cells and increases electrical efficiency of the PV panel by 8.3%.

In Chapter 7, Timilsina and colleagues analyze the technical, economic, and policy aspects of solar energy development and deployment. While the cost of solar energy has declined rapidly in the recent past, it still remains much higher than the cost of conventional energy technologies. Like other renewable energy technologies, solar energy benefits from fiscal and regulatory incentives and mandates, including tax credits and exemptions, feed-in-tariff, preferential interest rates, renewable portfolio standards and voluntary green power programs in many countries. Potential expansion of carbon credit markets also would provide additional incentives to solar energy deployment; however, the scale of incentives provided by the existing carbon market instruments, such as the Clean Development Mechanism of the Kyoto Protocol, is limited. Despite the huge technical potential, development and large-scale, market-driven deployment of solar energy technologies world-wide still has to overcome a number of technical and financial barriers. Unless these barriers are overcome, maintaining and increasing electricity supplies from solar energy will require continuation of potentially costly policy supports.

In Chapter 8, Yi and colleagues carry out a reliability based economical assessment of large-scale PV systems utilizing Universal Generating Function (UGF) techniques. The reliability models of solar panel arrays, PV inverters and energy production units (EPUs) are represented as the corresponding UGFs. The expected energy production models for different PV system configurations have were developed. The expected unit cost of electricity was calculated to provide informative metrics for making optimal decisions. The authors apply this method to determine the PV system configuration which provides electricity for a water purification process.

Chapter 9, by Denholm and Hummon, evaluates the operation of Concentrating Solar Power (CSP) with Thermal Energy Storage (TES) in two scenarios of renewable penetration in a test system based on two balancing areas in Colorado and Wyoming. The authors find that the simulated CSP plants were dispatched to avoid the highest-cost generation, generally shifting energy production to the morning and evening in non-summer months and shifting energy towards the end of the day in summer months. This shifting minimized the overall system production cost by reducing use of the least-efficient gas generators or preferentially displacing combined cycle generation over coal generation. The system also dispatches

CSP during the periods of highest net load, resulting in a very high capacity value. Overall, the authors conclude that the addition of TES to CSP increases its value; however, the difference in value between plants with and without storage is highly dependent on both the cost of natural gas and the penetration of other renewable sources, such as PV.

In Chapter 10, Banoni and colleagues examine the cost and benefits, both financial and environmental, of two forms of solar power generation, grid-tied photovoltaic cells, and Dish Stirling Systems, using conventional carbon-based fuel as a benchmark. This chapter shows that both technologies are a sensible investment for consumers, but given that the dish Stirling consumer receives 6.37 dollars per watt while the home photovoltaic system consumer receives between 0.9 and 1.70 dollars per watt, the former appears to be a superior option. Despite the large investment, this chapter deduces that it is far more feasible to get few strong investors to develop a solar farm of large installed capacity, than to get thousands of households to install photovoltaic arrays in their roofs. Potential implications of the solar farm construction include an environmental impact given the size of land require for this endeavour. However, the positive aspects, which include a large CO_2 emission reduction aggregated over the lifespan of the farm, outweigh any minor concerns or potential externalities.

In Chapter 11, Farhar and colleagues examine the social acceptance of utility-scale concentrating solar power project. The authors focus on social factors that may facilitate and impede the adoption and implementation of CSP—a technology that captures the sun's thermal energy using curved mirrors to focus sunlight onto a high temperature receiver. Based on a case study, the authors suggested set of policies, if implemented, can significantly improve community acceptance of solar energy projects.

Hampton and Eckermann show in Chapter 12 that social learning principles can provide a range of benefits for communication and decision making in the informed promotion of grid-connected photovoltaic technology. Public perceptions and citizens' investment decisions should move beyond framing decisions relative to subsidy levels to consider long-term investment returns. Retailers and installers of residential photovoltaic systems are encouraged to promote the option of building-integrated panels given favourable preferences shown for such technology.

Muyiwa Adaramola

PART I

OVERVIEW

CHAPTER 1

HYBRID SOLAR: A REVIEW ON PHOTOVOLTAIC AND THERMAL POWER INTEGRATION

T. T. CHOW, G. N. TIWARI, AND C. MENEZO

1.1 INTRODUCTION

In the past 3–4 decades, the market of solar thermal and photovoltaic (PV) electricity generation has been growing rapidly. So were the technological developments in hybrid solar photovoltaic/thermal (PVT) collectors and the associated systems. Generally speaking, a PVT system integrates photovoltaic and solar thermal systems for the co-generation of electrical and thermal power from solar energy. A range of methods are available such as the choices of monocrystalline/polycrystalline/amorphous silicon (c-Si/pc-Si/a-Si) or thin-film solar cells, air/liquid/evaporative collectors, flat-plate/concentrator types, glazed/unglazed designs, natural/forced fluid flow, and stand-alone/building-integrated features. Accordingly, the systems are ranging from PVT air and/or water heating system to hot-water supply through PV-integrated heat pump/pipe or combined heating and cooling and to actively cooled PV concentrator through the use of lens/ reflectors. Engineering considerations can be on the selection of heat removal fluid, the collector type, the balance of system, the thermal to electrical yield ratio, the solar fraction, and so on. These all have determining effects on the system operating mode, working temperature, and energy performance.

Theoretical and experimental studies of PVT were documented as early as in mid 1970s [1–3]. Despite the fact that the technical validity was early concluded, only in recent years that it has gained wide attention. The amount of publications grows rapidly. The following gives an overview of the development of the technology, placing emphasis on the research and development activities in the last decade. Readers may refer to Chow [4] for a better understanding of the early developments.

1.2 PVT DEVELOPMENTS IN THE TWENTIETH CENTURY

1.2.1 EARLY WORKS ON COLLECTOR DESIGN

The early research works were mainly on flat-plate collectors [5, 6]. Garg and his coworkers carried out mathematical and experimental studies on PVT systems [7–9]. Sopian et al. developed steady-state models, for comparing the performance of single- and doublepass PVT/a collectors [10, 11]. Through transient analysis, Prakash [12] pointed out that the air collector (PVT/a) design is lower in thermal efficiency than the water collector (PVT/w), because of the inferior heat transfer between the thermal absorber and the airflow stream. Bergene and Løvvik [13] derived a detailed physical model of a flat-plate PVT/w collector, through which the total efficiency was evaluated.

de Vries [14] investigated the performance of several PVT collector designs. The single-glazed design was found better than the unglazed (of which the thermal efficiency is unfavorable) or the double-glazed design (of which the electrical efficiency is unfavorable). Nevertheless, exergy analysis performed by Fujisawa and Tani [15] indicated that the exergy output density of the unglazed design is slightly higher than the single-glazed option, taking the fact that the thermal energy contains more unavailable energy. For low temperature water heating applications like for swimming pool-water heating, the unglazed PVT/w system is recommended. In cold winter days, antifreeze liquid can be used but then the summer performance will be affected [16].

Rockendorf et al. [17] compared the performance of a thermoelectric collector (first generating heat and subsequently electricity) and a PVT/w collector (in sheet-and-tube design); the electrical output of the PVT/w collector was found significantly higher than the thermoelectric collector.

In the above mathematical and experimental studies, the reported thermal efficiency of practical PVT/liquid systems is generally in the range of 45 to 70% for unglazed to glazed collector designs. For flat-plate PVT/a systems, the optimal thermal efficiency can be up to 55%.

1.2.2 DEVELOPMENTS TOWARDS COMPLEX SYSTEMS

In the 1990s, the initiative of PVT research was apparently a response to the global environmental deterioration and the growing interest in building-integrated photovoltaic (BiPV) designs. Comparing with the separated PV systems, the building integration of PV modules improves the overall performance and durability of the building facade. Nevertheless, building integration may bring the cell temperatures up to 20°C above the normal working temperature [18]. Other than the benefits of cooling, PVT collectors provide aesthetical uniformity than the side-by-side arrays of PV and solar thermal collectors. Alternative cooling schemes of the BiPV systems were examined [19–21]. Hollick [22] assessed the improvement in the system energy efficiency when solar cells were added onto the solar thermal metallic cladding panels on vertical facades.

Continued successfulness on concentrator-type (c-PVT) systems began to take shape. Akbarzadeh and Wadowski [23] studied a heat-pipe-based coolant design which is a linear, trough-like system. Luque et al. [24] successfully developed a concentrating array using reflecting optics and one-axis tracking. By that time, facing the conflicting roles of water heating and PV cooling, the design temperature of water that leaves a PVT/w collector is not high. Combining PVT and solar-assisted heat pump (SAHP) technology was then seen as a good alternative. Ito et al.

[25] constructed a PVT-SAHP system with pc-Si aluminum roll-bond solar panels.

Generally speaking, in the 20th century the PVT research works had been mostly focused on improving the cost-performance ratio as compared to the solar thermal and PV systems installed side by side. For real-building projects the PVT/a systems were more readily adopted in Europe and North America, though the higher efficiency of the PVT/w system has been confirmed by that time. Solar houses with PVT/w provision were once sold in Japan in late 1990s. Unfortunately such innovative housing was in lack of demand in the commercial market [26]. A summary of the PVT technology in the period, including the marketing potentials, was reported by the Swiss Federal Office [27] and the International Energy Agency (IEA) [28].

1.3 RECENT DEVELOPMENTS IN FLAT-PLATE PVT

1.3.1 PVT AIR COLLECTOR SYSTEMS

1.3.1.1 COLLECTOR DESIGN AND PERFORMANCE

The PVT air collectors, either glazed or unglazed, provide simple and economical solution to PV cooling. The air can be heated to different temperature levels through forced or natural flow. Forced circulation is more effective than natural circulation owing to better thermal convective and conductive behavior, but the fan power consumption reduces the net electricity output. Their use is mostly to meet the demands on industrial hot air, indoor space heating, and/or agricultural dehydration.

Hegazy investigated the thermal, electrical, hydrodynamic, and overall performance of four types of flat-plate PVT/a collectors [29]. These included channel above PV as Mode 1, channel below PV as Mode 2, PV between single-pass channels as Mode 3, and finally the double-pass design as Mode 4. The numerical analysis showed that while Mode 1 has the lowest performance, the other three have comparable energy outputs. On the whole, Mode 3 requires the least fan power.

Tripanagnostopoulos et al. carried out outdoors tests on different PVT/a and PVT/w collector configurations in Patra, Greece [30]. It was suggested to place the collectors in parallel rows and keeping a distance between adjacent rows to avoid shading. Diffuse reflectors then were placed between the adjacent rows to enlarge the received radiation at collector surfaces. Their experimental tests at noon hour gave a range of thermal efficiency from 38% to 75% for PVT/a collectors and 55% to 80% for PVT/w designs, depending on whether the reflectors were in place. The research team [31] further studied numerically the effect of adding suspended metal sheet at the middle of the air channel and the finned arrangements at the opposite wall of the air channel. It was found that such low cost improvements are more relevant to small collector length and can be readily applied to BiPVT/a installations. They [32] also introduced a PVT/bi-fluid collector incorporated with improvements identified in their previous work.

Tiwari et al. explored the overall efficiency performance and optimal designs of an unglazed PVT/a collector [33]. Energy matrices were derived considering the embodied energy at different processing stages in India [34]. Raman and Tiwari [35] then studied the annual thermal and exergy efficiencies of their proposed PVT/a collector for five different climate zones. The exergy efficiency was found unfavorable under strong solar radiation. Also the double-pass design shows better performance than the single-pass option; this echoes the findings of Sopian et al. [10] and Hegazy [29]. Furthermore, the life cycle analysis showed that the energy payback time (EPBT) in India is about 2 years. Also evaluated were the effect of fill factor [36] and the integrated performance with an earth air heat exchanger system [37]. Further, Dubey et al. [38] compared different configurations of glass-to-glass and glass-to-tedlar PV modules in Delhi. Experiments found that the glass-to-glass module is able to achieve higher supply air temperature and electrical efficiency. Their study extended to derive the analytical expressions for multiple PVT/a collectors connected in series, including the testing procedures [39, 40].

Assoa et al. in France introduced a PVT/bi-fluid collector that integrates preheating and domestic hot-water production [41]. The design includes alternate positioning of the solar thermal collector section and the PV section. The higher fluid temperature output allows the flexibility such as coupling with solar cooling devices during summer and facilitates a direct domestic hot-water system without adding auxiliary heating device. Parametric studies showed that the thermal efficiency could reach 80% under favorable collector length and mass flow rate conditions.

Sukamongkol et al. [42] studied the dynamic performance of a condenser desiccant for air conditioning energy reduction with the use of double-pass PVT/a collector. The thermal energy generated by the system was able to produce warm dry air as high as 53°C and 23% relative humidity. Electricity of about 6% of the daily total solar radiation can be obtained. Moreover, together with the heat recovery from the condenser to regenerate the desiccant for dehumidification, around 18% of the air conditioning energy can be saved.

Ali et al. [43] investigated the characteristics of convective heat transfer and fluid flow inside a PVT/a channel with the provision of a single row of oblique plates array. These plates arrays were positioned obliquely to the flow direction with variable oblique angles and with separations that avoid the partial shading of solar cells. The study was initiated taking the fact that the entrance region of a heated fluid flow channel is characterized by differentiating thermal and hydrodynamics boundary layers; the convective heat-transfer coefficient is then substantially larger than that at downstream locations. Thus, using oblique (interrupted) plates in a duct, or a channel, to prevent fully developed flow formation has the advantage of obtaining enhanced heat-transfer characteristics.

Kumar and Rosen [44] investigated the effect of adding vertical fins to the lower air channel of a double-pass PVT/a collector. The extended fin area was found able to reduce the cell temperature significantly.

1.3.1.2 BUILDING-INTEGRATED OPTIONS (BIPVT/A)

In conventional BiPV systems, an air gap is often provided at the rear of the PV arrays for the air cooling of modules by natural convection. The heat recovery from the air stream for a meaningful use constitutes a BiPVT/a system. From a holistic viewpoint, Bazilian and Prasad [45] summarized its potential applications. The multifunctional façade or roof was ideal for PVT integration that produces heat, light, and electricity simultaneously, in addition to the building shelter functionality.

(1) Works in Europe

In UK, the Brockstill Environment Centre in Leicester opened in 2001 was equipped with a roof-mounted PVT/a system [46]. To assess the performance of various operational and control modes, a combined simulation approach was adopted with the use of two popular thermal simulation tools: ESP-r and TRNSYS. Monitored actual energy use data of the building shows very positive results.

Mei et al. [47] studied the dynamic performance of a BiPVT/a collector system constructed in the 90s at the Mataro Library in Spain. Their TRNSYS model was validated against experimental data from a pc-Si PV facade. The heating and cooling loads for various European buildings with and without such a ventilated facade were then evaluated. The simulation results showed that more winter heating energy can be saved for the use of the preheated ventilation in a building located in Barcelona, but less is for Stuttgart in Germany and Loughborough in UK. The higher latitude locations therefore need a higher percentage of solar air collectors in the combined system. Further, Infield et al. [48] explored different approaches to estimate the thermal performance of BiPVT/a facades, including a design methodology based on an extension of the familiar heat loss and radiation gain factors.

The main difficulty in analyzing BiPVT/a performance lies in the prediction of its thermal behavior. When the temperature profile and the sun shading situation are known, the electrical performance can be readily determined. This is not the case for thermal computation. The estimation of the convective heat-transfer coefficients, for example, is far from direct. The actual processes may involve a mix of forced and natural convection, laminar and turbulent flow, and, simultaneously, the developing flow at the air entrance. The external wind load on the panels further complicates the situation. For a semitransparent facade, thermal energy enters and transmits through the air cavity both directly (for glazing transmission) and indirectly (through convection and radiation exchange). The heat transfer to the ventilating stream is probably most complex, particularly for buoyant flow.

Sandberg and Moshfegh derived analytical expressions for the coolant flow rate, velocity, and temperature rise along the length of the vertical channel behind the PV panels [49]. Their experimental results were well matching the theoretical predictions for constrained flow, but were less accurate for ducts with opened ends. For the latter, Mittelman et al. developed a generalized correlation for the average channel Nusselt number for the combined convective-radiative cooling [50]. Their solution of the governing equations and boundary conditions was computed through CFD analysis. Gan also studied the effect of channel size on the PV performance through CFD analysis [51]. To reduce possible overheating or hot spot formation, the required minimum air gaps were determined. Experimental works on a PVT façade were undertaken by Zogou and Stapountzis [52] for better understanding of the flow and turbulence with natural and forced convection modes. Supported by CFD modeling, the results show that the selection of flow rate and the heat-transfer characteristics of the back sheet are critical.

(2) Works in North America

In Canada, Chen et al. [53, 54] introduced a BiPVT/a system to a near net-zero energy solar house in Eastman Quebec. The solar house, built in 2007, featured with ventilated concrete slabs (VCSs). A VCS is a type of forced-air thermoactive building systems in which the concrete slabs exchange thermal energy with the air passage through its internal hollow voids. The BiPVT system is designed to cover one continuous roof surface to enhance aesthetic appeal and water proofing. Outdoor air is drawn by a variable speed fan with supervisory control to achieve the desired supply temperature. On a sunny winter day, the typical air temperature rise was measured 30–35°C. The typical thermal efficiency was at least 20% based on the gross roof area. Analysis of the monitored data showed that the VCS was able to accumulate thermal energy during a series of clear sunny days without overheating the slab surface or the living space.

Athienitis et al. [55] presents a design concept with transpired collector. This was applied to a full-scale office building demonstration project in Montreal. The experimental prototype was constructed with UTC (open-loop unglazed transpired collector) of which 70% surface area was covered with black-frame PV modules specially designed to enhance solar energy absorption and heat recovery. The system was compared side by side with a UTC of the same area under outdoor sunny conditions with low wind. This project was considered a near optimal application in an urban location in view of the highly favorable system design. While the thermal efficiency of the UTC system was found higher than the BiPVT/a combined thermal plus electrical efficiency, the equivalent thermal efficiency of the BiPVT/a system (assuming that electricity can be converted to four times as much heat) can be 7–17% higher.

Pantic et al. [56] compared 3 different open-loop systems via mathematical models. These include Configuration 1: unglazed BiPVT roof, Configuration 2: unglazed BiPVT roof connected to a glazed solar air collector, and Configuration 3: glazed BiPVT. It was pointed out that air flow in the BiPVT cavity should be selected as a function of desired outlet temperatures and fan energy consumption. Cavity depths, air velocity in the air cavity, and wind speed were found having significant effect on the unglazed BiPVT system energy performance. Development of efficient fan control strategies has been suggested an important step. Configurations 2 and 3 may be utilized to significantly increase thermal efficiency and air outlet temperature. In contrast, Configuration 3 significantly reduces electricity production and may lead to excessive cell temperatures and is thus not recommended unless effective means for heat removal are in place. The unglazed BIPVT system linked to a short vertical solar air collector is suitable for a connection with a rock bed thermal storage.

(3) Works in Asia Pacific

For warm climate applications, the ventilated BiPV designs are found better than the PVT/a designs with heat recovery. Crawford et al. [57] compared the EPBT of a conventional c-Si BiPV system in Sydney with two BiPVT/a systems with c-Si and a-Si solar cells, respectively. They found that the EPBT of the above three installations are in the range of 12–16.5 years, 4–9 years, and 6–14 years, respectively. The two BiPVT/a options reduce the EPBT to nearly one-half.

Agrawal and Tiwari [58, 59] studied a BiPVT/a system on the rooftop of a building, under the cold climatic conditions of India. It is concluded that for a constant mass flow rate of air, the series connected collectors are more suitable for the building fitted with the BIPVT/a system as rooftop.

For a constant velocity of air flow, the parallel combination is then the better choice. While the c-Si BiPVT/a systems have higher energy and exergy efficiencies, the a-Si BiPVT systems are the better options from the economic point of view.

Jie et al. [60] studied numerically the energy performance of a ventilated BiPV façade in Hong Kong. It was found that the free airflow gap affects little the electrical performance, but is able to reduce the heat transmission through the PV façade. Yang et al. [61] carried out a similar study based on the weather conditions of three cities in China: Hong Kong, Shanghai, and Beijing. It was found that on typical days the ratio of space cooling load reduction owing to the ventilated PV facade is 33–52%.

Chow et al. [62] investigated the BiPVT/a options of a hotel building in Macau, with the PVT facade associated with a 24-hour air-conditioned room. The effectiveness of PV cooling by means of natural airflow was investigated with two options: free openings at all sides of the air gap as Case 1 and in Case 2 the enclosed air gap that behaves as a solar chimney for air preheating. These were also compared with the conventional BiPV without ventilation. The ESP-r simulation results showed an insignificant difference in electricity output from the three options. This was caused by a reverse down flow at the air gap at night, owing to the cooling effect of a 24-hour air-conditioned room located behind the PVT facade. It was concluded that both the climate condition and system operating mode affect significantly the PV productivity.

In China, Ji et al. [63] studied theoretically and experimentally the performance of a photovoltaic-Trombe wall, which was constructed at an outdoor environmental chamber. This south-facing façade in Hefei was composed of a PV glazing (with pc-Si cells) at the outside and an insulation wall at the inside with top and bottom vent openings. This leaves a natural flow air channel in between for space heating purpose. The results confirmed its dual benefits—improving the room thermal condition (with 5–7°C air temperature rise in winter) and generating electricity (with cell efficiency at 10.4% on average).

(4) Works on Window Systems

In Sweden, a multifunction PVT hybrid solar window was proposed by Fieber et al. [64]. The solar window is composed of thermal absorbers on which PV cells are laminated. The absorbers are building integrated into the inside of a standard window, thus saving frames and glazing and also the construction cost. Reflectors are placed behind the absorbers for reducing the quantity of cells. Via computer simulation, the annual electrical output shows the important role of diffuse radiation, which accounts for about 40% of the total electricity generation. Compared to a flat PV module on vertical wall, this solar window produces about 35% more electrical energy per unit cell area.

Vertical collectors and windows are more energy efficient at high-latitude locations, considering the sun path. Davidsson et al. [65] studied the performance of the above hybrid solar window in Lund, Sweden (55.44°N). Also a full-scale system combining four of these solar windows was constructed in a single family home in Alvkarleo, Sweden (60.57°N). The solar window system was equipped with a PV-driven DC pump. The projected solar altitude is high in summer, and accordingly a large portion of the solar beam falls directly onto the absorber with a minor contribution from the reflector. This is the ideal operating mode of the solar window, with the reflector partly opened and the window delivers heat, electricity, and light altogether. Effects of different control strategies for the position of the rotatable reflector were also studied, so was the performance comparison with roof collector [66].

A ventilated PV glazing consists of a PV outer glazing and a clear inner glazing. The different combinations of vent openings allow different modes of ventilating flow, which can be buoyant/induced or mechanical/driven. The space heating mode belongs to the BiPVT/a category. Besides the popularly used opaque c-Si solar cells on glass, the see-through a-Si solar window can also be used. Chow et al. [67] analyzed its application in the office environment of Hong Kong. The surface transmissions were found dominated by the inner glass properties. The overall heat transfer however is affected by both the outer and inner glass properties. Experimental comparisons were made between the use of PV glazing and absorptive glazing [68]. The comparative study on single, double, and double-ventilated cases showed that the ventilated PV glazing is able to reduce the direct solar gain and glare effectively. The savings on air-conditioning electricity consumption are 26% for the single-glazing case and 82% for the ventilated double-glazing case. Further, via a validated ESP-r simulation model [69], the natural-ventilated PV technology was found reducing the air-conditioning power consumption by 28%, comparing with the conventional single absorptive glazing system. With daylight control, additional saving in artificial lighting can be enhanced [70].

1.3.2 PVT LIQUID COLLECTOR SYSTEMS

1.3.2.1 PVT/W COLLECTORS

(1) Collector Design and Applications
Zondag et al. compared the energy performance of different PVT/w collector design configurations [71, 72]. The efficiency curves of nine collector configurations were obtained through computer analysis. At zero reduced temperature, the thermal efficiencies of the unglazed and single-glazed sheet-and-tube collectors were found 52% and 58%, respectively, and that of the channel-above-PV design is 65%. Also compared were the annual yields when these collectors were assumed to serve a DHW system. The channel-below-PV (transparent) configuration was found having the highest overall efficiency. Nevertheless, the more economical single-glazed sheet-and-tube design was recommended for DHW production since its efficiency was found only 2% less. For low-temperature water heating, the unglazed PVT/w collector is recommended.

Sandnes and Rekstad developed a PVT/w collector with c-Si solar cells pasted on polymer thermal absorber [73]. Square-shape box-type absorber channels were filled with ceramic granulates. This improves heat transfer to flowing water. The opposite surface was in black color which allows it to serve as a solar thermal collector when turned up-side-down. The analysis showed that the presence of solar cells reduces the heat absorption by about 10% of the incident radiation, and the glazing (if exists) reduces the optical efficiency by around 5%. It was expected to serve well in low-temperature water-heating system.

Chow introduced an explicit dynamic model for analyzing transient performance of single-glazed sheet-and-tube collector [74]. Through the multinodal finite different scheme, the dynamic influences of intermittent solar irradiance and autocontrol device operation can be readily investigated. The appropriateness of the nodal scheme was evaluated through sensitivity tests. The study also reveals the importance of having good thermal contact between the water tubing and the thermal absorber, as well as between the absorber and the encapsulated solar cells.

Zakharchenko et al. also pointed out the importance of good thermal contact between solar cells and thermal absorber [75]. So the direct use of commercial PV module in PVT collectors is not recommended. They introduced a substrate material with 2 mm aluminum plate covered by 2 µm insulating film, of which the thermal conductivity was only 15% less than that of aluminum. They also pointed out that the solar cell area should be smaller than the size of the absorber and should be at the portion of the collector where the coolant enters. As an echo to this last point, Dubey and Tiwari [76] examined the performance of a self-sustained single-glazed PVT/w collector system with a partial coverage of PV module (packing factor = 0.25) in Delhi. The electricity generated from the PV module positioned at the water inlet end was used to drive a DC pump.

Kalogirou [77] developed a TRNSYS model of a pump-operated domestic PVT/w system complete with water tank, power storage and conversion, and temperature differential control. Further, Kalogirou and Tripanagnostopoulos [78] examined domestic PVT/w applications working with either thermosyphon or pump circulation modes. Their simulation study covered 12 cases with pc-Si and a-Si PV modules, and in three cities: Athens in Greece, Nicosia in Cyprus, and Madison in USA. The results showed that the economical advantage is more obvious for Nicosia and Athens where the availability of solar radiation is higher. Similar conclusions can be reached when comparing comparable applications at an industrial scale [79]. Also in Cyprus, Erdil et al. [80] carried out experimental measurements on an open-loop PVT/w domestic water-preheating system. Water flowed by gravity into a channel-above-PV type collector. The CPBT was estimated around 1.7 years.

Vokas et al. [81] performed a theoretical analysis of PVT/w application in domestic heating and cooling systems in three cities that belong to different climate zones, namely, Athens, Heraklion, and Thessaloniki. The thermal efficiency was found around 9% lower than the conventional solar thermal collector. Hence the interpolation of the PV laminate only affects slightly the thermal efficiency. The difference between the mentioned two systems in the percentage of domestic heating and cooling load coverage is only around 7%.

The effect of reflectors on PVT/w collector equipped with c-Si solar cells was studied by Kostić et al. [82]. Both numerical computation and experimental measurements arrived at the same optimal angle positions of the bottom reflector. The results show the positive effect of reflectors made of aluminum sheet and, considering the additional cost of about 10% for the reflectors, there is an energy gain in the range of 20.5–35.7% in summer.

Saitoh et al. [83] carried out the experimental study of a single-glazed sheet-and-tube PVT collector using brine (propylene glycol) solution as the coolant. Field measurements at a low energy house in Hokkaido were also observed. With a solar fraction of 46.3%, the system electrical efficiency was 8-9% and thermal efficiency 25–28%. When compared with the conventional system, the payback periods were found 2.1 years for energy, 0.9 years for GHG emission, and 35.2 years for cash flow, respectively.

The use of optimized working fluid (like nanofluid) was proposed through a numerical study by Zhao et al. [84]. The system consists of a PV module using c-Si solar cell and a thermal unit based on the direct absorption collector (DAC) concept. First the working fluid of the thermal unit absorbs the solar infrared radiation. Then, the remaining visible light is transmitted and converted into electricity by the solar cell. The arrangement prevents the excessive heating of the solar cell. The system works for both nonconcentrated and concentrated solar radiation. The optical properties of the working fluid were optimized to maximize the transmittance and the absorptance of the thermal unit in the visible and infrared part of the spectrum, respectively.

Chow et al. compared the performance of glazed and unglazed sheet-and-tube thermosyphon PVT/w collector systems in Hong Kong through theoretical models as well as experimental tests [85]. The evaluation indicates that the glazed design is always suitable if either the thermal or the overall energy output is to be maximized, but the exergy analysis supports the use of unglazed design if the increase of PV cell efficiency, packing factor, ratio of water mass to collector area, and wind velocity are seen as the desirable factors. Similar experimental work was done by J. H. Kim and J. T. Kim in Korea [86]; the results show that the thermal efficiency of the glazed collector is 14% higher than the unglazed alternative, but the unglazed one had electrical efficiency 1.4% higher than the glazed design. Further for the unglazed option, they compared the performance of the conventional sheet-and-tube thermal absorber with the rectangular-box-channel design, which was made of aluminum. At zero reduced temperature, the thermal and electrical efficiencies were found 66% and 14%, respectively, whereas those of the box-channel configuration were 70% and 15%, respectively [87].

Dubey and Tiwari [88] analyzed the thermal energy, exergy, and electrical energy yield of PVT/w sheet-and-tube collectors in India. Based on a theoretical model, the number of collectors in use, their series/parallel connection patterns, and the weather conditions were examined. For enhancing economical/environmental benefits, the optimum hot-water withdrawal rate was evaluated [89]. Optimum PVT/w system configuration was also evaluated by Naewngerndee et al. [90] via CFD employing the finite element method.

Rosa-Clot et al. [91] suggested a PVT configuration with water flow in polycarbonate box above the PV panel. The water layer absorbs the infrared radiation leaving the visible part almost unaffected. Efficiencies were evaluated and in particular the effects of temperature and irradiance mismatching on PV outputs were discussed.

(2) Absorber Materials

In view of the limitations on the fin performance of a sheet-and-tube PVT/w collector [74], an aluminum-alloy box-channel PVT/w collector was developed through the collaborative efforts of the City University of Hong Kong and the University of Science and Technology of China. Several generations of the collector prototypes were produced and tested under the subtropical Hong Kong and temperate Hefei climatic conditions [92–95]. The thermosyphon system was found working well in both locations. Dynamic simulations showed that better convective heat transfer between the coolant and the channel wall can be achieved by reducing the channel depth and increasing the number of channels per unit width [95]. Sensitivity tests in Hefei showed that the daily cell efficiency reaches

10.2%, daily primary energy saving efficiency reaches 65% with a packing factor of 0.63 [96]. In Hong Kong, the CPBT was found to be 12 years which is comparable to the more bulky side-by-side arrangement and is much better than the 52 years for plain PV module operation [97].

Affolter et al. [98] pointed out that the typical solar performances of PVT/liquid collectors are similar to those of nonselective-type solar thermal absorbers. Observations showed that the stagnation temperature (i.e., the elevated panel temperature in the absence of water flow) of the absorber of a solar thermal collector with a state of-the art spectrally selective coating may reach 220°C. Since a PVT absorber generally has higher solar reflectance and higher infrared emission than a solar thermal absorber, the stagnation temperature may be lowered to 150°C. But this is still higher than 135°C; that is, the temperature that the common encapsulation materials like EVA (ethylene vinyl acetate) resin may withstand [99]. EVA oxidizes rapidly at above 135°C.

Charalambous et al. [100] carried out a mathematical analysis on the optimum copper absorber plate configuration having the least material content and thus cost, whilst maintaining high collector efficiency. Both header-and-riser arrangement and serpentine arrangement were studied. It was found that light weight collector design can be achieved using very thin fins and small tubes.

The possible use of copolymer absorber to replace the commonly used metallic sheet-and-tube absorber had been examined extensively [101, 102]. This replacement offers several advantages:

1. the weight reduction leads to less material utilization and easier installation;
2. the manufacturing process is simplified since fewer components are involved;
3. the above leads to a reduction in production costs.

However, there are disadvantages such as low thermal conductivity, large thermal expansion, and limited service temperature. On the other hand, the copolymer in use has to be good in physical strength, UV light protected, and chemically stable.

Huang et al. studied a PVT/w collector system complete with DC circulating pump and storage tank [103]. The collector was fabricated by the attachment of commercial PV modules on a corrugated polycarbonate absorber plate with square-shaped box channels.

Cristofari et al. studied the performance of a PVT/w collector with polycarbonate absorber and pc-Si PV modules carrying top and bottom glass sheets [104]. Water in forced flow passed through parallel square channels at very low flow rate and so with negligible pumping power. The system design capacity was based on the hot-water demands for the inhabitants at Ajaccio in France. With the use of a mathematical model, the annual averaged efficiencies of 55.5% for thermal, 12.7% for PV, 68.2% for overall, and 88.8% for energy saving were obtained. The maximum stagnation temperature at the absorber was found 116.2°C, which is acceptable. They further developed a collector with copolymer material that reduces the weight by more than half in comparison with the conventional metallic one [105].

Fraisse et al. suggested that PVT/liquid system is very suitable for the low temperature operation of Direct Solar Floor (DSF) system [106]. An application example in the Macon area of France was evaluated with the use of a glazed collector system. With propylene glycol as the coolant, the TRNSYS simulation results gave the annual c-Si cell efficiency as 6.8%, that is, a 28% drop as compared to a conventional nonintegrated PV module. Without the front glazing, the cell efficiency was increased to 10% as a result of efficient cooling. It was also found that, in the case of a glazed collector with a conventional control system for DSF, the maximum temperature at the PV modules was above 100°C in summer. At this temperature level, the use of EVA in PV modules will be subject to strong risks of degradation. The use of either a-Si cells or unglazed collector was recommended.

(3) PVT Collector Design

Santbergen et al. [107] carried out a numerical study on a forced-flow PVT/w system. Single-glazed sheet-and-tube flat-plate PVT collectors were employed and designed for grid-connected PV system with c-Si PUM cells. Both the annual electrical and thermal efficiencies were found around 15% lower, when compared to separate conventional PV and conventional solar thermal collector systems. It was suggested that both the electrical and the thermal efficiency can be improved through the use of antireflective coatings. Alternatively, the thermal efficiency can be improved by the application of low-e coating, but at the expenses of the electrical efficiency.

Since long wavelength irradiance with photon energies below the bandgap energy is hardly absorbed at all, the solar absorptance of the solar cells is significantly lower than that of a black absorber (with absorptance = 0.95). Santbergen and van Zolingen [108] also suggested two methods to increase long wavelength absorption:

1. to use semitransparent solar cells followed by a second absorber and
2. to increase the amount of long wavelength irradiance absorption in the back contact of the solar cell.

Computer analysis showed that these two methods are able to achieve an overall absorption of 0.87 and 0.85, respectively.

Dupeyrat et al. [109] developed a PV cell lamination with Fluorinated Ethylene Propylene (FEP) at the front. This results in an alternative encapsulation with a lower refractive index than glass pane and a lower UV absorbing layer than conventional EVA material. Experimental tests showed an increase of more than $2 \, mA/cm^2$ in generated current density for the PVT module. Finally the developments led to a new covered PVT collector for domestic hot-water application [110]. The c-Si PV cells were directly laminated on an optimized aluminium heat exchanger. The thermal efficiency at zero-reduced temperature was measured 79% with a corresponding electrical efficiency of 8.8%, leading to a high overall efficiency of almost 88%. This PVT collector in the standard conditions is therefore reaching the highest efficiency level reported in the literature.

Employing a bifacial PV module having two active surfaces can to generate more electric power than the traditional one-surface module. The optical properties of water allow its absorption of light mainly in the infrared region. This is compatible with PV modules using shorter wavelengths in the solar spectra for its electricity conversion. The water absorption only slightly affects the working region of a-Si PV cell (decrease of water transparency at around 950 nm), but it strongly absorbs the light with wavelengths above 1100 nm (the "thermal part" of the solar spectrum). Therefore, a PVT/w collector system with Si bifacial solar PV module can be advantageous. In Mexico, Robles-Ocampo et al. [111] carried out experimental test on a PVT/w system with c-Si bifacial PV module in Queretaro. The transparent flat collector was fabricated with a 15 mm channel underneath a glass cover, which was found better than the plastic cover in terms of service life. Stainless steel mirror reflectors (to prevent oxidation in the outdoor environment) were used for illuminating the rear face of the solar cells. Measurements found that the glass water-filled flat collector placed above the PV module reduces the front face efficiency by 10%. When considering the radiation flux incident directly onto the active elements of the hybrid system, the system is able to achieve an electrical efficiency around 16% and an equivalent thermal efficiency around 50%.

1.3.2.2 BUILDING-INTEGRATED SYSTEMS (BIPVT/W)

The research works on BiPVT/w systems have been less popular than the BiPVT/a systems. Ji et al. carried out a numerical study of the annual performance of a BiPVT/w collector system for use in the residential buildings of Hong Kong [112]. Pump energy was neglected. Assuming perfect bonding of PV encapsulation and copper tubing onto the absorber, the annual thermal efficiencies on the west-facing facade were found 47.6% and 43.2% for film cells and c-Si cells, respectively, and the cell efficiencies were 4.3% and 10.3%. The reductions in space heat gain were estimated 53.0% and 59.2%, respectively.

Chow et al. studied a BiPVT/w system applicable to multistory apartment building in Hong Kong [113]. The TRNSYS system simulation program was used. They also constructed an experimental BiPVT/w system at a rooftop environmental chamber [114]. The energy efficiencies of thermosyphon and pump circulation modes were compared across the subtropical summer and winter periods. The results show the better energy performance of the thermosyphon operation, with thermal efficiency reaches 39% at zero-reduced temperature and the corresponding cell efficiency 8.6%. The space cooling load is reduced by 50% in peak summer. Ji et al. [115] further carried out an optimization study on this type of installation. The appropriate water flow rate, packing factor and connecting pipe diameter were determined.

Based on the above-measured data, Chow et al. also developed an explicit dynamic thermal model of the BiPVT/w collector system [116]. Its annual system performance in Hong Kong reconfirmed the better performance of the natural circulation mode. This is because of the elimination of the pumping power and hence better cost saving [117]. The CPBT was 13.8 years, which is comparable to the stand-alone box channel PVT/w collector system. This BiPVT/w application is able to shorten the CPBT to one-third of the plain BiPV application. The corresponding energy payback time (EPBT) and greenhouse-gas payback time (GPBT) were found 3.8 years and 4.0 years [118]; these are much more favorable than CPBT.

Anderson et al. analyzed the design of a roof-mounted BiPVT/w system [119]. Their BiPVT/w collector prototype was integrated to the standing seam or toughed sheet roof, on which passageways were added to the trough for liquid coolant flow. Their modified Hottel-Whillier model was validated experimentally. The results showed that the key design parameters, like fin efficiency, lamination requirements, and thermal conductivity between the PV module and the supporting structure, affect significantly the electrical and thermal efficiencies. They also suggested that a lower cost material like precoated steel can replace copper or aluminum for thermal absorption since this does not significantly reduce the efficiencies. Another suggestion was to integrate the system "into" (rather than "onto") the roof structure, as the rear air space in the attic can provide a high level of thermal insulation. The effect of nonuniform water flow distribution on electrical conversion performance of BiPVT/w collector of various size was studied by Ghani et al. [120]. The numerical work identified the important role of the array geometry.

Eicker and Dalibard [121] studied the provision of both electrical and cooling energy for buildings. The cooling energy can be used for the direct cooling of activated floors or ceilings. Experimental works with uncovered PVT collector prototypes were carried out to validate a simulation model, which then calculated the night radiative heat exchange with the sky. Large PVT frameless modules were then developed and implemented in a residential zero energy building and tested.

Matuska compared the performance of two types of fin configurations of BiPVT/w collector systems with the BiPV installation using pc-Si cells [122]. Two different European climates and for roof/façade applications were evaluated by computer simulation. Better energy production potential of the BiPVT/w collector systems was confirmed—the results show 15–25% increase in electricity production in warm climate (Athens) and 8–15% increase in moderate climate (Prague). The heat production by steady flow forced convection can be up to 10 times higher than the electricity production.

Corbin and Zhai [123] monitored a prototype full-scale BiPVT/w collector installed on the roof of a residential dwelling. Measured performance was used to develop a CFD model which was subsequently used in a parametric study to assess the collector performance under a variety of operating conditions. Water temperature observed during testing reaches 57.4°C at an ambient temperature of 35.3°C. The proposed BiPVT/w collector shows a potential for providing the increased electrical efficiency of up to 5.3% above a naturally ventilated BiPV roof.

1.3.3 PVT REFRIGERATION

1.3.3.1 HEAT-PUMP INTEGRATION (PVT/HEAT PUMP)

Conventional air-to-air heat pumps cannot function efficiently in cold winter with extreme low outdoor air temperatures. Bakker et al. [124] introduced a space and tap-water heating system with the use of roof-sized PVT/w array combined with a ground coupled heat pump. The system performance, as applied to one-family Dutch dwelling, was evaluated through TRNSYS simulation. The results showed that the system is able to satisfy all heating demands, and at the same time, to meet nearly all of its electricity consumption, and to keep the long-term average ground temperature constant. The PVT system also requires less roof space and offers architectural uniformity while the required investment is comparable to those of the conventional provisions.

Bai et al. [125] presented a simulation study of using PVT/w collectors as water preheating devices of a solar-assisted heat pump (SAHP) system. The system was for application in sports center for swimming pool heating and also for bathroom services. The energy performances of the same system under different climatic conditions, that included Hong Kong and three other cities in France, were analyzed and compared. Economic implications were also determined. The results show that although the system performance in Hong Kong is better than the cities in France, the cost payback period is the longest in Hong Kong since there was no government tax reduction.

Extensive research on PVT/heat pump system with variable pump speed has been conducted in China. Experimental investigations were performed on unglazed PVT evaporator system prototype [126, 127]. Mathematical models based on the distributed parameters approach were developed and validated [128, 129]. The simulation results show that its performance can be better than the conventional SAHP system. With R-134a as the refrigerant, the PV-SAHP system is able to achieve an annual average COP of 5.93 and PV efficiency 12.1% [130].

In the warm seasons, glazed PVT collector may not serve well as PVT evaporator. In cold winter however, the outdoor temperature can be much lower than the evaporating temperature of the refrigeration cycle. Then the heat loss at the PV evaporator is no longer negligible. The front cover would be able to improve both the photothermic efficiency and the system COP. Pei et al. concluded that for winter operation, the overall PVT exergy efficiency as well as the COP can be improved in the presence of the glass cover [131]. This is beneficial since the space heating demand is higher in winter.

1.3.3.2 PVT-INTEGRATED HEAT PIPE

These works were basically done in China. Based on the concept of integrating heat pipes and a PVT flat-plate collector into a single unit, Pei et al. [132, 133] designed and constructed an experimental rig of heat-pipe PVT (HP-PVT) collector system. The HP-PVT collector can be used in cold regions without freezing, and corrosion can be reduced as well. The evaporator section of the heat pipes is connected to the back of the aluminum absorber plate, and the condenser section is inserted into a water box above the absorber plate. The PV cells are laminated onto the surface of the aluminum plate. Detailed simulation models were developed and validated by the experimental findings. Through these, parametric analyses as well as annual system performance for use in three typical climatic areas in China were predicted. The results show that for the HP-PVT system without auxiliary heating equipment, in Hong Kong there are 172 days a year that the hot water can be heated to more than 45°C using solar energy. In Lhasa and Beijing, the results are 178 days and 158 days for the same system operation.

In order to solve the nonuniform cooling of solar PV cells and control the operating temperature of solar PV cells conveniently, Wu et al. [134] developed a heat-pipe PVT hybrid system by selecting a wick heat pipe to absorb isothermally the excessive heat from solar cells. The PV modules were in a rectangular arrangement, and below which the wick heat-pipe evaporator section is closely attached. The thermal-electric conversion performance was theoretically investigated.

1.3.3.3 PVT TRIGENERATION

Calise et al. [135] studied the possible integration of medium-temperature and high-temperature PVT collectors with solar heating and cooling technology, and hence a polygeneration system that produces electricity, space heating and cooling, and domestic hot water. A case study was performed with PVT collectors, single-stage absorption chiller, storage tanks, and auxiliary heaters as the main system components. The system performance was analyzed from both energetic and economic points of view. The economic results show that the system under investigation in Italy can be profitable, provided that an appropriate funding policy is available.

1.4 RECENT DEVELOPMENTS IN CONCENTRATOR-TYPE DESIGN

The use of concentrator-type PVT (or c-PVT) collector can to increase the intensity of solar radiation on the PV cells than the flat-plate collector. The c-PVT collectors are generally classified into three groups: single cells, linear geometry, and densely packed modules. Higher efficiency solar cells that handle higher current can be used, although they are more expensive than the flat-plate module cells. The complex sun tracking driving mechanism also incurs additional costs [136]. But the benefit is that a considerable portion of the cell surfaces can be replaced by low-cost reflector surfaces. Connecting the solar cells in series can to increase the output voltage and decrease the current at a given power output. This reduces the ohmic losses. During operation, nonuniform temperature can exist across the cells. The cell at the highest temperature will limit the efficiency of the whole string [137]. Hence the c-PVT coolant circuit should be designed to keep the cell temperature uniform and relatively low. A precise shape of the reflector surface and an accurate alignment is also essential, particularly when the concentration ratio is high. A precise tracking system is also important.

Refractive lenses and reflector surfaces are commonly in use in c-PVT. Comparatively, lens is lower in weight and material costs. For systems designed for higher concentration, more concentrator material per unit cell/absorber area is in need. Then the use of lenses is more appropriate. However, concentrator systems that utilize lenses are unable to focus scattered light. This limits their usage to places with mostly clear weather. On the other hand, using liquid as the coolant is more effective than using air to obtain better electrical output. These make reflector-type c-PVT systems good for medium- to high-temperature hot-water systems that are required for cooling, desalination, or other industrial processes. At lower operating temperatures, a flat-plate collector may have higher efficiency than the c-PVT collector when both are directly facing the sun. But at higher temperature differential, the large exposed surface of a flat-plate collector leads to more thermal loss. So the performance gap between the two will diminish when the working temperature gradually increases.

Rosell et al. in Spain constructed a low-concentrating PVT prototype with the combination of flat-plate channel-below-PV (opaque) collector and linear Fresnel concentrator that worked on two-axis tracking system [138]. The total efficiency was found above 60% when the concentration ratio was above 6x. Their theoretical analysis reconfirms the importance of the cell-absorber thermal conduction.

Experimental trough c-PVT systems with energy flux ratio in the range of 10–20 were developed and tested in China by Li et al. [139]. Performances of arrays with the use of different solar cells types were compared. Ji et al. [140] also developed steady models of the system and validated them by the measured data. They found that the system performance can be optimized by improving the mirror reflectivity and the thermal solar radiation absorptivity of the lighting plate and by pursuing a suitable focal line with uniform light intensity distribution. Also as a China-UK joint research effort, a CPC-based PVT system with a U-pipe was investigated [141]. CPC stands for compound parabolic concentrator. The U-pipe avoids the temperature gradient on the whole absorber and on every block cell and simultaneously produces electricity using the same temperatures. More recently, Zhang et al. [142] proposed a PV system with integrated CPC plate that adopts a low precision solar tracking method; the performance can be better than the fixed installation or the case with periodic adjustment only in several months.

Coventry developed a combined heat and power solar (CHAPS) collector system in Australia [143]. This was a linear trough system designed for single tracking. The c-Si solar cells (at 20% standard conversion efficiency) in row were bonded to an aluminum receiver and were cooled by water with antifreeze and anticorrosive additives flowing in an internally finned aluminum pipe. Light was focused onto the cells through the use of glass-on-metal parabolic reflectors (92% reflectance) and at high concentration ratio (37x). Under typical operating conditions the measurements gave a thermal efficiency around 58%, electrical efficiency around 11%, and a combined efficiency around 69%.

Kribus et al. [144] developed a miniature concentrating PV system that can be installed on any rooftop. The design is based on a small parabolic dish which is similar to a satellite dish. The system equipments are relatively easy to deliver and handle without the use of special tools. By concentrating sunlight about 500 times, the solar cell area is greatly reduced.

In high-latitude countries like Sweden, the solar radiation is asymmetric over the year because of the high cloud coverage during winter, and thus concentrated to a small angular interval of high irradiation. This makes the use of economical stationary reflectors or concentrators attractive. Cost reduction can be realized by laminating thin aluminum foil on steel substrate. Nilsson et al. [145] carried out experimental tests on an asymmetric compound parabolic reflector system, with two different truncated parabolic reflectors made of anodized aluminum and aluminum laminated steel, respectively. Their measurements confirmed that changing the back reflector from anodized aluminum to aluminum laminated steel does not change the energy output. They also found that the optimal cell position is to face the front reflector, assuming no space restriction. This will result in the lowest cost for electricity generation. For cases with limited roof space, they suggested to place the solar cells on both sides of the absorber. This considers that, once a trough with cells on one side of the absorber is constructed, the cost of adding cells to the other side is relatively low.

A two-stage hybrid device was theoretically studied by Vorobiev et al. [146, 147], with solar cells incorporated on energy flux concentrator and heat-to-electric/mechanic energy converter. Two option cases were investigated:

1. system with the separation of "thermal solar radiation", and
2. system without solar spectrum division and solar cell operating at high temperature.

The first case allows the solar cell to operate at a low ambient temperature, but then requires the production of a new kind of solar cell which does not absorb or dissipate solar radiation as infrared. The calculations showed that with a concentration as high as 1500x, the total conversion efficiency could reach 35–40%. The solar cell in the second option is subject to concentrated sunlight. It was found that with the use of GaAs-based single-junction cell having room temperature efficiency at 24% and a concentrator at 50x, the total conversion efficiency is around 25–30%. If a higher concentration is used, the efficiency can be even higher.

Jiang et al. [148] introduced a two-stage parabolic trough concentrating PVT system, which contains a concentrator, a spectral beam splitting filter, an evacuated collector tube, and the solar cell components. The nondimensional optical model with the focal length of the concentrator as the characteristic length has been developed to analyze the properties of the concentrating system using the beam splitting filter. The geometry concentration ratio and the size of solar image at different structure parameters have been obtained. It is shown that using the filter the heat load of the cell can be reduced by 20.7%. Up to 10.5% of the total incident solar energy can be recovered by the receiver, and the overall optical efficiency in theory is about 0.764.

Kostic et al. [149] studied the influence of reflectance from flat-plate solar radiation concentrators made of aluminum sheet and aluminum foil on energy efficiency of PVT collector. The total reflectance from concentrators made of aluminum sheet and aluminum foil is almost the same, but specular reflectance (which is bigger in concentrators made of aluminum foil) results in an increase of solar radiation intensity concentration factor. The total energy generated by c-PVT collector made of aluminum foil in optimal position is higher than the total energy generated by those made of aluminum sheet.

The basic feature of an STPV (solar thermophotovoltaic) is in the use of high temperature emitter as an intermediate element that absorbs concentrated solar light and emits photonic energy to solar cells through which the thermal radiation energy is converted to electricity. Compared with the solar cells, the STPV system can utilize the concentrated solar energy sufficiently. It conveniently adjusts the spectral feature of photons released from the emitter corresponding to the bandgaps of solar cells in the system by controlling the emitter temperature and/or installing the spectral filter. Xuan et al. [150] established the design and optimization method of STPV systems by taking into account the energy transport and/or conversion processes among the solar concentrator, the emitter, the spectral filter, the solar cells, and the cooling subsystem. The effects of the nonparallelism of sun rays, aperture ratios, and the tracking error on concentration capacity were investigated. The emitters made of different materials and with different configurations were numerically analyzed. The effects of concentration ratio, spectral characteristic of the filter, series and shunt resistance of the cell, and the performance of the cooling system on the STPV systems were discussed. Compared with the one-dimensional photonic filer, the optimized nonperiodic filter has a better performance. A high-performance cooling system is required to keep the cell temperature below 50°C.

As an attempt to improve the system efficiency of concentrating photovoltaics (CPVs), an investigation has been done by Kosmadakis et al. [151] into the technical aspects as well as the cost analysis, by combining the technologies of the CPV and the organic Rankine cycles (ORCs). The heat rejected from the CPV is recovered from the ORC, in order to increase the total electric power output. The findings constitute evidence that the CPV-ORC system can be an alternative for recovering the heat from concentrating PVs. Nevertheless, the mechanical power produced from the expander of the ORC can be used in other applications as well.

Huang et al. [152] suggested a PVT system based on organic photovoltaics (OPVs). The OPV cells were fabricated onto one-half of a tubular light pipe inside which the silicone oil was flowed. This allows solar energy in the visible wavelengths to be effectively converted into electricity by photocell while simultaneously the silicone oil captures the infrared radiation part of the spectrum as heat energy. The oil filled tube acts as a passive optical element that concentrates the light into the PV and thereby increases its overall efficiency.

While silicon-based PV technology has many physical barriers, it is expected that the future PVT developments will be closely linked to the breakthroughs in solar cell technology. The next generation solar cells such as polymer, nanocrystalline, and dye-sensitized solar cells will be less expensive, flexible, compact, lightweight, and efficient. Take dye-sensitized solar cells (DSSCs) as an example, the operation does not need the p-n junction but mimics the principle of natural photosynthesis. It is composed of a porous layer of titanium dioxide nanoparticles, covered with a molecular dye that absorbs sunlight, like the chlorophyll in green leaves. The DSSCs today convert about 11 to 12% of the sunlight into electricity. The use of hybrid ZnO/TiO_2 photoanodes will be able to utilize the high electron transport rate of ZnO and the high electron injection efficiency and stability of TiO_2 materials [153].

1.5 MISCELLANEOUS DEVELOPMENTS IN RECENT YEARS

1.5.1 AUTONOMOUS APPLICATIONS

Desalination is a process to produce the distilled water from brackish/saline water by means of solar still. Solar distillation of brackish water is a good option to obtain fresh water in view of its simple technology and low energy operation. A proposed design of PVT-integrated active solar still was tested in India by Kumar and Tiwari [154–156]. This PVT active solar still is self-sustainable and can be used in remote areas. Compared with a passive solar still, the daily distillate yield was found 3.5 times higher, and 43% of the pumping power can be saved. Based on 0.05 m water depth, the range of CPBT can be shortened from 3.3–23.9 years to 1.1–6.2 years (depending on the selling price of distilled water) and the EPBT from 4.7 years to 2.9 years. The hybrid active solar still is able to provide higher electrical and overall thermal efficiency, which is about 20% higher than the passive solar still. On the other hand, Gaur and Tiwari [157] conducted a numerical study to optimize the number of collectors for PVT/w hybrid active solar still. The number of PVT collectors connected in series has been integrated with the basin of a solar still.

Another potential application lies in crop drying, which is the process of removing excess moisture from crop produced through evaporation, either by natural or forced convection mode. Tiwari et al. developed a PVT mixed mode dryer together with an analytical model for performance analysis [158]. The experimental tests were executed for the forced convection mode under no load conditions. The annual gains for different Indian cities were evaluated and the results show that Jodhpur is the best place for the installation of this type of PVT dryer.

1.5.2 HIGH TEMPERATURE APPLICATIONS

Mittelman et al. [159] studied the application of c-PVT system in a LiBr absorption chiller designed for single effect. In the theoretical analysis, the desorber inlet temperature was set in the range of 65–120°C and without thermal storage. The PV module was based on triple-junction cells with a nominal conversion efficiency of 37%. A typical dish concentrator with an 85% optical efficiency was used. The results showed that the loss in cell efficiency owing to the increase in operation temperature was insignificant. Under a reasonably range of economic conditions, the c-PVT cooling system can be comparable to, and sometimes even better than, a conventional cooling system.

A c-PVT water desalination system was also proposed by Mittelman et al. [160], in which a c-PVT collector field was to couple to a large-scale multiple-effect evaporation thermal desalination system. Small dish concentrator type was used in the numerical analysis. The vapor formed in each evaporator condenses in the next (lower temperature) effect and thus provides the heat source for further evaporation. Additional feed preheating is to be provided by vapor process bleeding from each effect. The range of top brine temperature is from 60 to 80°C. Through numerical analysis, this approach was found competitive relative to other solar-driven desalination systems and even relative to the conventional reverse-osmosis desalination. Because of the higher ratio of electricity to heat generation, the high concentration option with the use of advanced solar cells can be advantageous.

1.5.3 COMMERCIAL ASPECTS

The commercial markets for both solar thermal and photovoltaic are growing rapidly. It is expected that the PVT products, once become mature, would experience a similar trend of growth. In future, the market share might be even larger than that for solar thermal collectors. The higher energy output characteristics of the PVT collector suit better the increasing demands on low-energy or even zero-carbon buildings. Nevertheless, although there are plenty reported literatures on the theoretical and experimental findings of PVT collector systems, those reporting on full-scale application and long-term monitoring have been scarce [161]. The number of commercial systems in practical services remains small. The majority involves flat-plate collectors but only with limited service life. The operating experiences are scattered. In the inventory of IEA Solar Heating and Cooling Task 35, over 50 PVT projects have been identified in the past 20 years. Less than twenty of these projects belong to the PVT/w category which is supposed to have better application potential. On the other hand, while most projects were in Europe such as UK and Netherlands, there have been projects realized in Thailand, in which large-scale glazed a-Si PVT/w systems were installed at hospital and government buildings [162]. It is important to have full documentation of the initial testing and commissioning, as well as the long-term monitoring of the real systems performance, including the operating experiences and the problems encountered. Developments in the balance of system are also important—for example the improvement works in power quality and power factor in PV inverter design [163]. The improvements in power supply stability with power conditioner and better integration of renewable energy sources on to utility grid have been other key research areas [164].

Standard testing procedures for PVT commercial products are so far incomplete. In essence, the performance of PVT commercial products can be tested either outdoor or indoor. The outdoor test needs to be executed in steady conditions of fine weather, which should be around noon hours and preferably with clear sky and no wind. This can be infrequent; say for Northern Europe, it may take six months to acquire the efficiency curve [165]. Indoor test can be quicker and provides repeatable results. To make available an internationally accepted testing standard is one important step for promoting the PVT products.

Although there have been an obvious increase in academic publications in hybrid PVT technology in recent years, many key issues related to the commercialization of PVT products are still not resolved. The lack of economic viability, public awareness, product standardization, warranties and performance certification, installation training, and experiences are the barriers. It is important for the reliability of the technology to be thoroughly assessed.

1.6 CONCLUSION

Global climate change and fuel supply security have led to the fast development in renewable technology, including solar energy applications. The installations of solar thermal and PV electricity generation devices are growing rapidly and these lead to an increase in the demand of PVT collector system. PVT products have much shorter CPBT than the PV counterpart. Hence PVT (rather than PV) as a renewable energy technology is expected to first become competitive with the conventional power generating systems.

In the past decades, the performance of various PVT collector types had been studied theoretically, numerically, and experimentally. This paper serves to review the endeavor in the past years. While in the early works the research efforts were on the consolidation of the conceptual ideas and the feasibility study on basic PVT collector designs, the PVT studies from the '90s onward have been more related to the collector design improvement and economical/environmental performance evaluation. There were more rigorous numerical analyses of the energy and fluid flow phenomena on conventional collectors with an experimental validation. The ideas of building-integrated design emerged and the demonstration projects were reported. Since the turn of century, the focus has been generally shifted towards the development of complimentary products, innovative systems, testing procedures, and design optimization. The marketing potential and justification on various collector designs and system applications have been evaluated through user feedback, life cycle cost, and/or embodied energy evaluations. The computational analyses become more comprehensive with the use of powerful analytical tools. There have been increased uses of explicit dynamic modeling techniques and also public domain simulation programs, including CFD codes. The evaluation has been extended to geographical comparison of long-term performance based on typical year round weather data on one hand and the second-law thermodynamic assessment on the other. International research collaborations and related activities have been increasing.

Despite the sharp increase in academic activities, the developments of commercial products and real system applications are still limited. The issues of investment costs and product reliability are to be fully attended. More efforts must be on the identification of suitable product materials, manufacturing techniques, testing and training requirements, potential customers, market strength, and so on.

REFERENCES

3.   M. Wolf, "Performance analyses of combined heating and photovoltaic power systems for residences," Energy Conversion, vol. 16, no. 1-2, pp. 79–90, 1976.

4.   E. C. Kern and M. C. Russell, "Combined photovoltaic and thermal hybrid collector systems," in Proceedings of the 13th IEEE Photovoltaic Specialists, pp. 1153–1157, Washington DC, USA, June 1978.

5.   L. W. Florschuetz, "Extension of the Hottel-Whillier model to the analysis of combined photovoltaic/thermal flat plate collectors," Solar Energy, vol. 22, no. 4, pp. 361–366, 1979.

6. T. T. Chow, "A review on photovoltaic/thermal hybrid solar technology," Applied Energy, vol. 87, no. 2, pp. 365–379, 2010.
7. C. H. Cox III and P. Raghuraman, "Design considerations for flat-plate-photovoltaic/thermal collectors," Solar Energy, vol. 35, no. 3, pp. 227–241, 1985.
8. B. Lalović, Z. Kiss, and H. Weakliem, "A hybrid amorphous silicon photovoltaic and thermal solar collector," Solar Cells, vol. 19, no. 2, pp. 131–138, 1986.
9. H. P. Garg and R. S. Adhikari, "Conventional hybrid photovoltaic/thermal (PV/T) air heating collectors: steady-state simulation," Renewable Energy, vol. 11, no. 3, pp. 363–385, 1997.
10. H. P. Garg and R. S. Adhikari, "Optical design calculations for CPCS," Energy, vol. 23, no. 10, pp. 907–909, 1998.
11. H. P. Garg and R. S. Adhikari, "Performance analysis of a hybrid photovoltaic/thermal (PV/T) collector with integrated CPC troughs," International Journal of Energy Research, vol. 23, no. 15, pp. 1295–1304, 1999.
12. K. Sopian, K. S. Yigit, H. T. Liu, S. Kakaç, and T. N. Veziroglu, "Performance analysis of photovoltaic thermal air heaters," Energy Conversion and Management, vol. 37, no. 11, pp. 1657–1670, 1996.
13. K. Sopian, H. T. Liu, S. Kakac, and T. N. Veziroglu, "Performance of a double pass photovoltaic thermal solar collector suitable for solar drying systems," Energy Conversion and Management, vol. 41, no. 4, pp. 353–365, 2000.
14. J. Prakash, "Transient analysis of a photovoltaic-thermal solar collector for co-generation of electricity and hot air/water," Energy Conversion and Management, vol. 35, no. 11, pp. 967–972, 1994.
15. T. Bergene and O. M. Løvvik, "Model calculations on a flat-plate solar heat collector with integrated solar cells," Solar Energy, vol. 55, no. 6, pp. 453–462, 1995.
16. D. W. de Vries, Design of a photovoltaic/thermal combi-panel [Ph.D. thesis], Eindhoven Technical University, Eindhoven, The Netherlands, 1998.
17. T. Fujisawa and T. Tani, "Annual exergy evaluation on photovoltaic-thermal hybrid collector," Solar Energy Materials and Solar Cells, vol. 47, no. 1–4, pp. 135–148, 1997.
18. B. Norton and J. E. J. Edmonds, "Aqueous propylene-glycol concentrations for the freeze protection of thermosyphon solar energy water heaters," Solar Energy, vol. 47, no. 5, pp. 375–382, 1991.
19. G. Rockendorf, R. Sillmann, L. Podlowski, and B. Litzenburger, "PV-hybrid and thermoelectric collectors," Solar Energy, vol. 67, no. 4–6, pp. 227–237, 1999.
20. M. W. Davis, A. H. Fanney, and B. P. Dougherty, "Prediction of building integrated photovoltaic cell temperatures," Journal of Solar Energy Engineering, vol. 123, no. 3, pp. 200–210, 2001.
21. B. Moshfegh and M. Sandberg, "Flow and heat transfer in the air gap behind photovoltaic panels," Renewable and Sustainable Energy Reviews, vol. 2, no. 3, pp. 287–301, 1998.
22. J. A. Clarke, J. W. Hand, C. M. Johnstone, N. Kelly, and P. A. Strachan, "Photovoltaic-integrated building facades," Renewable Energy, vol. 8, no. 1–4, pp. 475–479, 1996.
23. B. J. Brinkworth, R. H. Marshall, and Z. Ibarahim, "A validated model of naturally ventilated PV cladding," Solar Energy, vol. 69, no. 1, pp. 67–81, 2000.

24. J. C. Hollick, "Solar cogeneration panels," Renewable Energy, vol. 15, no. 1, pp. 195–200, 1998.

25. A. Akbarzadeh and T. Wadowski, "Heat pipe-based cooling systems for photovoltaic cells under concentrated solar radiation," Applied Thermal Engineering, vol. 16, no. 1, pp. 81–87, 1996.

26. A. Luque, G. Sala, J. C. Arboiro, T. Bruton, D. Cunningham, and N. Mason, "Some results of the EUCLIDES photovoltaic concentrator prototype," Progress in Photovoltaics: Research and Applications, vol. 5, no. 3, pp. 195–212, 1997.

27. S. Ito, N. Miura, and K. Wang, "Performance of a heat pump using direct expansion solar collectors," Solar Energy, vol. 65, no. 3, pp. 189–196, 1999.

28. S. Ito and N. Miura, "Photovoitaic and thermal hybrid systems," in Proceedings of the Asia-Pacific Conference of International Solar Energy Society, pp. 73–78, Gwangju, South Korea, October 2004.

29. P. Affolter, D. Ruoss, P. Toggweiler, and B. A. Haller, "New Generation of Hybrid Solar Collectors," Final Report DIS 56360/16868, Swiss Federal Office for Energy, 2000.

30. IEA, "Photovoltaic/thermal solar energy systems: status of the technology and roadmap for future development," Task 7 Report, International Energy Agency, PVPS T7-10, 2002.

31. A. A. Hegazy, "Comparative study of the performances of four photovoltaic/thermal solar air collectors," Energy Conversion and Management, vol. 41, no. 8, pp. 861–881, 2000.

32. Y. Tripanagnostopoulos, T. Nousia, M. Souliotis, and P. Yianoulis, "Hybrid photovoltaic/thermal solar systems," Solar Energy, vol. 72, no. 3, pp. 217–234, 2002.

33. J. K. Tonui and Y. Tripanagnostopoulos, "Air-cooled PV/T solar collectors with low cost performance improvements," Solar Energy, vol. 81, no. 4, pp. 498–511, 2007.

34. Y. Tripanagnostopoulos, "Aspects and improvements of hybrid photovoltaic/thermal solar energy systems," Solar Energy, vol. 81, no. 9, pp. 1117–1131, 2007.

35. A. Tiwari, M. S. Sodha, A. Chandra, and J. C. Joshi, "Performance evaluation of photovoltaic thermal solar air collector for composite climate of India," Solar Energy Materials and Solar Cells, vol. 90, no. 2, pp. 175–189, 2006.

36. A. Tiwari, P. Barnwal, G. S. Sandhu, and M. S. Sodha, "Energy metrics analysis of hybrid—photovoltaic (PV) modules," Applied Energy, vol. 86, no. 12, pp. 2615–2625, 2009.

37. V. Raman and G. N. Tiwari, "A comparison study of energy and exergy performance of a hybrid photovoltaic double-pass and single-pass air collector," International Journal of Energy Research, vol. 33, no. 6, pp. 605–617, 2009.

38. A. S. Joshi, I. Dincer, and B. V. Reddy, "Thermodynamic assessment of photovoltaic systems," Solar Energy, vol. 83, no. 8, pp. 1139–1149, 2009.

39. S. Nayak and G. N. Tiwari, "Energy metrics of photovoltaic/thermal and earth air heat exchanger integrated greenhouse for different climatic conditions of India," Applied Energy, vol. 87, no. 10, pp. 2984–2993, 2010.

40. S. Dubey, G. S. Sandhu, and G. N. Tiwari, "Analytical expression for electrical efficiency of PV/T hybrid air collector," Applied Energy, vol. 86, no. 5, pp. 697–705, 2009.

41. S. Dubey, S. C. Solanki, and A. Tiwari, "Energy and exergy analysis of PV/T air collectors connected in series," Energy and Buildings, vol. 41, no. 8, pp. 863–870, 2009.

42. S. C. Solanki, S. Dubey, and A. Tiwari, "Indoor simulation and testing of photovoltaic thermal (PV/T) air collectors," Applied Energy, vol. 86, no. 11, pp. 2421–2428, 2009.

43. Y. B. Assoa, C. Menezo, G. Fraisse, R. Yezou, and J. Brau, "Study of a new concept of photovoltaic-thermal hybrid collector," Solar Energy, vol. 81, no. 9, pp. 1132–1143, 2007.

44. Y. Sukamongkol, S. Chungpaibulpatana, B. Limmeechokchai, and P. Sripadungtham, "Condenser heat recovery with a PV/T air heating collector to regenerate desiccant for reducing energy use of an air conditioning room," Energy and Buildings, vol. 42, no. 3, pp. 315–325, 2010.

45. A. H. H. Ali, M. Ahmed, and M. S. Youssef, "Characteristics of heat transfer and fluid flow in a channel with single-row plates array oblique to flow direction for photovoltaic/thermal system," Energy, vol. 35, no. 9, pp. 3524–3534, 2010.

46. R. Kumar and M. A. Rosen, "Performance evaluation of a double pass PV/T solar air heater with and without fins," Applied Thermal Engineering, vol. 31, no. 8-9, pp. 1402–1410, 2011.

47. M. D. Bazilian and D. Prasad, "Modelling of a photovoltaic heat recovery system and its role in a design decision support tool for building professionals," Renewable Energy, vol. 27, no. 1, pp. 57–68, 2002.

48. B. P. Cartmell, N. J. Shankland, D. Fiala, and V. Hanby, "A multi-operational ventilated photovoltaic and solar air collector: application, simulation and initial monitoring feedback," Solar Energy, vol. 76, no. 1–3, pp. 45–53, 2004.

49. L. Mei, D. Infield, U. Eicker, and V. Fux, "Thermal modelling of a building with an integrated ventilated PV façade," Energy and Buildings, vol. 35, no. 6, pp. 605–617, 2003.

50. D. Infield, L. Mei, and U. Eicker, "Thermal performance estimation for ventilated PV facades," Solar Energy, vol. 76, no. 1–3, pp. 93–98, 2004.

51. M. Sandberg and B. Moshfegh, "Buoyancy-induced air flow in photovoltaic facades effect of geometry of the air gap and location of solar cell modules," Building and Environment, vol. 37, no. 3, pp. 211–218, 2002.

52. G. Mittelman, A. Alshare, and J. H. Davidson, "A model and heat transfer correlation for rooftop integrated photovoltaics with a passive air cooling channel," Solar Energy, vol. 83, no. 8, pp. 1150–1160, 2009.

53. G. Gan, "Effect of air gap on the performance of building-integrated photovoltaics," Energy, vol. 34, no. 7, pp. 913–921, 2009.

54. O. Zogou and H. Stapountzis, "Flow and heat transfer inside a PV/T collector for building application," Applied Energy, vol. 91, no. 1, pp. 103–115, 2012. ·

55. Y. Chen, A. K. Athienitis, and K. Galal, "Modeling, design and thermal performance of a BIPV/T system thermally coupled with a ventilated concrete slab in a low en-

ergy solar house: part 1, BIPV/T system and house energy concept," Solar Energy, vol. 84, no. 11, pp. 1892–1907, 2010.

56. Y. Chen, K. Galal, and A. K. Athienitis, "Modeling, design and thermal performance of a BIPV/T system thermally coupled with a ventilated concrete slab in a low energy solar house: part 2, ventilated concrete slab," Solar Energy, vol. 84, no. 11, pp. 1908–1919, 2010.

57. A. K. Athienitis, J. Bambara, B. O'Neill, and J. Faille, "A prototype photovoltaic/ thermal system integrated with transpired collector," Solar Energy, vol. 85, no. 1, pp. 139–153, 2011.

58. S. Pantic, L. Candanedo, and A. K. Athienitis, "Modeling of energy performance of a house with three configurations of building-integrated photovoltaic/thermal systems," Energy and Buildings, vol. 42, no. 10, pp. 1779–1789, 2010.

59. R. H. Crawford, G. J. Treloar, R. J. Fuller, and M. Bazilian, "Life-cycle energy analysis of building integrated photovoltaic systems (BiPVs) with heat recovery unit," Renewable and Sustainable Energy Reviews, vol. 10, no. 6, pp. 559–575, 2006.

60. B. Agrawal and G. N. Tiwari, "Optimizing the energy and exergy of building integrated photovoltaic thermal (BIPVT) systems under cold climatic conditions," Applied Energy, vol. 87, no. 2, pp. 417–426, 2010.

61. B. Agrawal and G. N. Tiwari, "Life cycle cost assessment of building integrated photovoltaic thermal (BIPVT) systems," Energy and Buildings, vol. 42, no. 9, pp. 1472–1481, 2010.

62. J. Jie, H. Wei, and H. N. Lam, "The annual analysis of the power output and heat gain of a PV-wall with different integration mode in Hong Kong," Solar Energy Materials and Solar Cells, vol. 71, no. 4, pp. 435–448, 2002.

63. H. Yang, J. Burnett, and Z. Zhu, "Building-integrated photovoltaics: effect on the cooling load component of building façades," Building Services Engineering Research and Technology, vol. 22, no. 3, pp. 157–165, 2001.

64. T. T. Chow, J. W. Hand, and P. A. Strachan, "Building-integrated photovoltaic and thermal applications in a subtropical hotel building," Applied Thermal Engineering, vol. 23, no. 16, pp. 2035–2049, 2003.

65. J. Ji, B. Jiang, H. Yi, T. T. Chow, W. He, and G. Pei, "An experimental and mathematical study of efforts of a novel photovoltaic-Trombe wall on a test room," International Journal of Energy Research, vol. 32, no. 6, pp. 531–542, 2008.

66. A. Fieber, H. Gajbert, H. Hakansson, J. Nilsson, T. Rosencrantz, and B. Karlsson, "Design, building integration and performance of a hybrid solar wall element," in Proceedings of the ISES Solar World Congress, Gothenburg, Sweden, 2003.

67. H. Davidsson, B. Perers, and B. Karlsson, "Performance of a multifunctional PV/T hybrid solar window," Solar Energy, vol. 84, no. 3, pp. 365–372, 2010.

68. H. Davidsson, B. Perers, and B. Karlsson, "System analysis of a multifunctional PV/T hybrid solar window," Solar Energy, vol. 86, no. 3, pp. 903–910, 2012. ·

69. T. T. Chow, K. F. Fong, W. He, Z. Lin, and A. L. S. Chan, "Performance evaluation of a PV ventilated window applying to office building of Hong Kong," Energy and Buildings, vol. 39, no. 6, pp. 643–650, 2007.

70. T. T. Chow, G. Pei, L. S. Chan, Z. Lin, and K. F. Fong, "A comparative study of PV glazing performance in warm climate," Indoor and Built Environment, vol. 18, no. 1, pp. 32–40, 2009.

71. T. T. Chow, Z. Qiu, and C. Li, "Potential application of "see-through" solar cells in ventilated glazing in Hong Kong," Solar Energy Materials and Solar Cells, vol. 93, no. 2, pp. 230–238, 2009.

72. T. T. Chow, Z. Qiu, and C. Li, "Performance evaluation of PV ventilated glazing," in Proceedings of the Building Simulation 2009, 11th International Building Performance Simulation Association Conference, Glasgow, UK, July 2009.

73. H. A. Zondag, D. W. de Vries, W. G. J. van Helden, R. J. C. van Zolingen, and A. A. van Steenhoven, "The thermal and electrical yield of a PV-thermal collector," Solar Energy, vol. 72, no. 2, pp. 113–128, 2002.

74. H. A. Zondag, D. W. de Vries, W. G. J. van Helden, R. J. C. van Zolingen, and A. A. van Steenhoven, "The yield of different combined PV-thermal collector designs," Solar Energy, vol. 74, no. 3, pp. 253–269, 2003.

75. B. Sandnes and J. Rekstad, "A photovoltaic/thermal (PV/T) collector with a polymer absorber plate. Experimental study and analytical model," Solar Energy, vol. 72, no. 1, pp. 63–73, 2002.

76. T. T. Chow, "Performance analysis of photovoltaic-thermal collector by explicit dynamic model," Solar Energy, vol. 75, no. 2, pp. 143–152, 2003.

77. R. Zakharchenko, L. Licea-Jiménez, S. A. Pérez-García et al., "Photovoltaic solar panel for a hybrid PV/Thermal system," Solar Energy Materials and Solar Cells, vol. 82, no. 1-2, pp. 253–261, 2004. ·

78. S. Dubey and G. N. Tiwari, "Thermal modeling of a combined system of photovoltaic thermal (PV/T) solar water heater," Solar Energy, vol. 82, no. 7, pp. 602–612, 2008.

79. S. A. Kalogirou, "Use a TRNSYS for modelling and simulation of a hybrid pv-thermal solar system for Cyprus," Renewable Energy, vol. 23, no. 2, pp. 247–260, 2001.

80. S. A. Kalogirou and Y. Tripanagnostopoulos, "Hybrid PV/T solar systems for domestic hot water and electricity production," Energy Conversion and Management, vol. 47, no. 18-19, pp. 3368–3382, 2006.

81. S. A. Kalogirou and Y. Tripanagnostopoulos, "Industrial application of PV/T solar energy systems," Applied Thermal Engineering, vol. 27, no. 8-9, pp. 1259–1270, 2007.

82. E. Erdil, M. Ilkan, and F. Egelioglu, "An experimental study on energy generation with a photovoltaic (PV)-solar thermal hybrid system," Energy, vol. 33, no. 8, pp. 1241–1245, 2008.

83. G. Vokas, N. Christandonis, and F. Skittides, "Hybrid photovoltaic-thermal systems for domestic heating and cooling—a theoretical approach," Solar Energy, vol. 80, no. 5, pp. 607–615, 2006.

84. L. T. Kostić, T. M. Pavlović, and Z. T. Pavlović, "Optimal design of orientation of PV/T collector with reflectors," Applied Energy, vol. 87, no. 10, pp. 3023–3029, 2010.

85. H. Saitoh, Y. Hamada, H. Kubota et al., "Field experiments and analyses on a hybrid solar collector," Applied Thermal Engineering, vol. 23, no. 16, pp. 2089–2105, 2003.

86. J. Zhao, Y. Song, W. H. Lam et al., "Solar radiation transfer and performance analysis of an optimum photovoltaic/thermal system," Energy Conversion and Management, vol. 52, no. 2, pp. 1343–1353, 2011.

87. T. T. Chow, G. Pei, K. F. Fong, Z. Lin, A. L. S. Chan, and J. Ji, "Energy and exergy analysis of photovoltaic-thermal collector with and without glass cover," Applied Energy, vol. 86, no. 3, pp. 310–316, 2009.

88. J. H. Kim and J. T. Kim, "Comparison of electrical and thermal performances of glazed and unglazed PVT collectors," International Journal of Photoenergy, vol. 2012, Article ID 957847, 7 pages, 2012. ·

89. J. H. Kim and J. T. Kim, "The experimental performance of an unglazed PVT collector with two different absorber types," International Journal of Photoenergy, vol. 2012, Article ID 312168, 6 pages, 2012. ·

90. S. Dubey and G. N. Tiwari, "Analysis of PV/T flat plate water collectors connected in series," Solar Energy, vol. 83, no. 9, pp. 1485–1498, 2009.

91. A. Tiwari, S. Dubey, G. S. Sandhu, M. S. Sodha, and S. I. Anwar, "Exergy analysis of integrated photovoltaic thermal solar water heater under constant flow rate and constant collection temperature modes," Applied Energy, vol. 86, no. 12, pp. 2592–2597, 2009.

92. R. Naewngerndee, E. Hattha, K. Chumpolrat, T. Sangkapes, J. Phongsitong, and S. Jaikla, "Finite element method for computational fluid dynamics to design photovoltaic thermal (PV/T) system configuration," Solar Energy Materials and Solar Cells, vol. 95, no. 1, pp. 390–393, 2011.

93. M. Rosa-Clot, P. Rosa-Clot, and G. M. Tina, "TESPI: thermal electric solar panel integration," Solar Energy, vol. 85, no. 10, pp. 2433–2442, 2011.

94. T. T. Chow, W. He, and J. Ji, "Hybrid photovoltaic-thermosyphon water heating system for residential application," Solar Energy, vol. 80, no. 3, pp. 298–306, 2006.

95. T. T. Chow, J. Ji, and W. He, "Photovoltaic-thermal collector system for domestic application," Journal of Solar Energy Engineering, vol. 129, no. 2, pp. 205–209, 2007.

96. W. He, T. T. Chow, J. Ji, J. Lu, G. Pei, and L. S. Chan, "Hybrid photovoltaic and thermal solar-collector designed for natural circulation of water," Applied Energy, vol. 83, no. 3, pp. 199–210, 2006.

97. J. Ji, J. Han, T. T. Chow, C. Han, J. Lu, and W. He, "Effect of flow channel dimensions on the performance of a box-frame photovoltaic/thermal collector," Proceedings of the Institution of Mechanical Engineers A, vol. 220, no. 7, pp. 681–688, 2006.

98. J. Ji, J. P. Lu, T. T. Chow, W. He, and G. Pei, "A sensitivity study of a hybrid photovoltaic/thermal water-heating system with natural circulation," Applied Energy, vol. 84, no. 2, pp. 222–237, 2007.

99. T. T. Chow, W. He, J. Ji, and A. L. S. Chan, "Performance evaluation of photovoltaic-thermosyphon system for subtropical climate application," Solar Energy, vol. 81, no. 1, pp. 123–130, 2007.

100. P. Affolter, W. Eisenmann, H. Fechner et al., PVT Roadmap: A European Guide for the Development and Market Introduction of PV-Thermal Technology, ECN, 2007.

101. H. A. Zondag and W. G. J. van Helden, "Stagnation temperature in PVT collector," Report ECN-RX-02-045, 2002, http://www.ecn.nl/library/reports/2002/rx02045. html.

102. P. G. Charalambous, S. A. Kalogirou, G. G. Maidment, and K. Yiakoumetti, "Optimization of the photovoltaic thermal (PV/T) collector absorber," Solar Energy, vol. 85, no. 5, pp. 871–880, 2011.

103. M. Meir and J. Rekstad, "Der solarnor kunststoffkollektor—the development of a polymer collector with glazing," in Proceedings of the Polymeric Solar Material, Erstes Leobener Symposium "Solartechnik—Neue moglichkeiten fur die Kunststoffbranche", pp. II. 1–II. 8., Polymer Competence Center Leoben, Leoben, Austria, October 2003.

104. P. Papillon, G. Wallner, and M. Kohl, "Polymeric materials in solar thermal applications," Solar Heating and Cooling Program, Task 39, International Energy Agency, http://www.iea-shc.org/task39.

105. B. J. Huang, T. H. Lin, W. C. Hung, and F. S. Sun, "Performance evaluation of solar photovoltaic/thermal systems," Solar Energy, vol. 70, no. 5, pp. 443–448, 2001.

106. C. Cristofari, G. Notton, and J. L. Canaletti, "Thermal behavior of a copolymer PV/Th solar system in low flow rate conditions," Solar Energy, vol. 83, no. 8, pp. 1123–1138, 2009.

107. C. Cristofari, J. L. Canaletti, G. Notton, and C. Darras, "Innovative patented PV/TH solar collector: optimization and performance evaluation," Energy Procedia, vol. 14, pp. 235–240, 2012. ·

108. G. Fraisse, C. Ménézo, and K. Johannes, "Energy performance of water hybrid PV/T collectors applied to combisystems of direct solar floor type," Solar Energy, vol. 81, no. 11, pp. 1426–1438, 2007.

109. R. Santbergen, C. C. M. Rindt, H. A. Zondag, and R. J. C. van Zolingen, "Detailed analysis of the energy yield of systems with covered sheet-and-tube PVT collectors," Solar Energy, vol. 84, no. 5, pp. 867–878, 2010.

110. R. Santbergen and R. J. C. van Zolingen, "Modeling the thermal absorption factor of photovoltaic/thermal combi-panels," Energy Conversion and Management, vol. 47, no. 20, pp. 3572–3581, 2006.

111. P. Dupeyrat, C. Ménézo, H. Wirth, and M. Rommel, "Improvement of PV module optical properties for PV-thermal hybrid collector application," Solar Energy Materials and Solar Cells, vol. 95, no. 8, pp. 2028–2036, 2011.

112. P. Dupeyrat, C. Ménézo, M. Rommel, and H. M. Henning, "Efficient single glazed flat plate photovoltaic-thermal hybrid collector for domestic hot water system," Solar Energy, vol. 85, no. 7, pp. 1457–1468, 2011.

113. B. Robles-Ocampo, E. Ruíz-Vasquez, H. Canseco-Sánchez et al., "Photovoltaic/thermal solar hybrid system with bifacial PV module and transparent plane collector," Solar Energy Materials and Solar Cells, vol. 91, no. 20, pp. 1966–1971, 2007.

114. J. Ji, T. T. Chow, and W. He, "Dynamic performance of hybrid photovoltaic/thermal collector wall in Hong Kong," Building and Environment, vol. 38, no. 11, pp. 1327–1334, 2003.

115. T. T. Chow, A. L. S. Chan, K. F. Fong, W. C. Lo, and C. L. Song, "Energy performance of a solar hybrid collector system in a multistory apartment building,"

Proceedings of the Institution of Mechanical Engineers A, vol. 219, no. 1, pp. 1–11, 2005.

116. T. T. Chow, W. He, and J. Ji, "An experimental study of façade-integrated photovoltaic/water-heating system," Applied Thermal Engineering, vol. 27, no. 1, pp. 37–45, 2007.

117. J. Ji, J. Han, T. T. Chow et al., "Effect of fluid flow and packing factor on energy performance of a wall-mounted hybrid photovoltaic/water-heating collector system," Energy and Buildings, vol. 38, no. 12, pp. 1380–1387, 2006.

118. T. T. Chow, W. He, A. L. S. Chan, K. F. Fong, Z. Lin, and J. Ji, "Computer modeling and experimental validation of a building-integrated photovoltaic and water heating system," Applied Thermal Engineering, vol. 28, no. 11-12, pp. 1356–1364, 2008.

119. T. T. Chow, A. L. S. Chan, K. F. Fong, Z. Lin, W. He, and J. Ji, "Annual performance of building-integrated photovoltaic/water-heating system for warm climate application," Applied Energy, vol. 86, no. 5, pp. 689–696, 2009.

120. T. T. Chow and J. Ji, "Environmental life-cycle analysis of hybrid solar photovoltaic/thermal systems for use in Hong Kong," International Journal of Photoenergy, vol. 2012, Article ID 101968, 9 pages, 2012. ·

121. T. N. Anderson, M. Duke, G. L. Morrison, and J. K. Carson, "Performance of a building integrated photovoltaic/thermal (BIPVT) solar collector," Solar Energy, vol. 83, no. 4, pp. 445–455, 2009.

122. F. Ghani, M. Duke, and J. K. Carson, "Effect of flow distribution on the photovoltaic performance of a building integrated photovoltaic/thermal (BIPV/T) collector," Solar Energy, vol. 86, no. 5, pp. 1518–1530, 2012. ·

123. U. Eicker and A. Dalibard, "Photovoltaic-thermal collectors for night radiative cooling of buildings," Solar Energy, vol. 85, no. 7, pp. 1322–1335, 2011.

124. T. Matuska, "Simulation study of building integrated solar liquid PV-T collectors," International Journal of Photoenergy, vol. 2012, Article ID 686393, 8 pages, 2012. ·

125. C. D. Corbin and Z. J. Zhai, "Experimental and numerical investigation on thermal and electrical performance of a building integrated photovoltaic-thermal collector system," Energy and Buildings, vol. 42, no. 1, pp. 76–82, 2010.

126. M. Bakker, H. A. Zondag, M. J. Elswijk, K. J. Strootman, and M. J. M. Jong, "Performance and costs of a roof-sized PV/thermal array combined with a ground coupled heat pump," Solar Energy, vol. 78, no. 2, pp. 331–339, 2005.

127. Y. Bai, T. T. Chow, C. Ménézo, and P. Dupeyrat, "Analysis of a hybrid PV/thermal solar-assisted heat pump system for sports center water heating application," International Journal of Photoenergy, vol. 2012, Article ID 265838, 13 pages, 2012. ·

128. J. Ji, K. Liu, T. T. Chow, G. Pei, W. He, and H. He, "Performance analysis of a photovoltaic heat pump," Applied Energy, vol. 85, no. 8, pp. 680–693, 2008.

129. J. Ji, G. Pei, T. T. Chow et al., "Experimental study of photovoltaic solar assisted heat pump system," Solar Energy, vol. 82, no. 1, pp. 43–52, 2008.

130. J. Ji, K. Liu, T. T. Chow, G. Pei, and H. He, "Thermal analysis of PV/T evaporator of a solar-assisted heat pump," International Journal of Energy Research, vol. 31, no. 5, pp. 525–545, 2007.

131. J. Ji, H. He, T. T. Chow, G. Pei, W. He, and K. Liu, "Distributed dynamic modeling and experimental study of PV evaporator in a PV/T solar-assisted heat pump," International Journal of Heat and Mass Transfer, vol. 52, no. 5-6, pp. 1365–1373, 2009.

132. T. T. Chow, K. F. Fong, G. Pei, J. Ji, and M. He, "Potential use of photovoltaic-integrated solar heat pump system in Hong Kong," Applied Thermal Engineering, vol. 30, no. 8-9, pp. 1066–1072, 2010.

133. G. Pei, J. Ji, T. T. Chow, H. He, K. Liu, and H. Yi, "Performance of the photovoltaic solar-assisted heat pump system with and without glass cover in winter: a comparative analysis," Proceedings of the Institution of Mechanical Engineers A, vol. 222, no. 2, pp. 179–187, 2008.

134. G. Pei, H. Fu, H. Zhu, and J. Ji, "Performance study and parametric analysis of a novel heat pipe PV/T system," Energy, vol. 37, no. 1, pp. 384–395, 2012. ·

135. G. Pei, H. Fu, J. Ji, T. T. Chow, and T. Zhang, "Annual analysis of heat pipe PV/T systems for domestic hot water and electricity production," Energy Conversion and Management, vol. 56, pp. 8–21, 2012. ·

136. S. Y. Wu, Q. L. Zhang, L. Xiao, and F. H. Guo, "A heat pipe photovoltaic/thermal (PV/T) hybrid system and its performance evaluation," Energy Build, vol. 43, no. 12, pp. 3558–3567, 2011. ·

137. F. Calise, M. D. d'Accadia, and L. Vanoli, "Design and dynamic simulation of a novel solar trigeneration system based on hybrid photovoltaic/thermal collectors (PVT)," Energy Conversion and Management, vol. 60, pp. 214–225, 2012. ·

138. A. Segal, M. Epstein, and A. Yogev, "Hybrid concentrated photovoltaic and thermal power conversion at different spectral bands," Solar Energy, vol. 76, no. 5, pp. 591–601, 2004.

139. A. Royne, C. J. Dey, and D. R. Mills, "Cooling of photovoltaic cells under concentrated illumination: a critical review," Solar Energy Materials and Solar Cells, vol. 86, no. 4, pp. 451–483, 2005.

140. J. I. Rosell, X. Vallverdu, M. A. Lechon, and M. Ibanez, "Design and simulation of a low concentrating photovoltaic/thermal system," Energy Conversion and Management, vol. 46, no. 18-19, pp. 3034–3046, 2005. ·

141. M. Li, G. L. Li, X. Ji, F. Yin, and L. Xu, "The performance analysis of the trough concentrating solar photovoltaic/thermal system," Energy Conversion and Management, vol. 52, no. 6, pp. 2378–2383, 2011.

142. X. Ji, M. Li, W. Lin, W. Wang, L. Wang, and X. Luo, "Modeling and characteristic parameters analysis of a trough concentrating photovoltaic/thermal system with GaAs and super cell arrays," International Journal of Photoenergy, vol. 2012, Article ID 782560, 10 pages, 2012. ·

143. G. Li, G. Pei, S. Yuehong, X. Zhou, and J. Ji, "Preliminary study based on building-integrated compound parabolic concentrators (CPC) PV/thermal technology," Energy Procedia, vol. 14, pp. 343–350, 2012. ·

144. L. Zhang, D. Jing, L. Zhao, J. Wei, and L. Guo, "Concentrating PV/T hybrid system for simultaneous electricity and usable heat generation: a review," International Journal of Photoenergy, vol. 2012, Article ID 869753, 8 pages, 2012. ·

145. J. S. Coventry, "Performance of a concentrating photovoltaic/thermal solar collector," Solar Energy, vol. 78, no. 2, pp. 211–222, 2005.

146. A. Kribus, D. Kaftori, G. Mittelman, A. Hirshfeld, Y. Flitsanov, and A. Dayan, "A miniature concentrating photovoltaic and thermal system," Energy Conversion and Management, vol. 47, no. 20, pp. 3582–3590, 2006.

147. J. Nilsson, H. Håkansson, and B. Karlsson, "Electrical and thermal characterization of a PV-CPC hybrid," Solar Energy, vol. 81, no. 7, pp. 917–928, 2007.
148. Y. Vorobiev, J. González-Hernández, P. Vorobiev, and L. Bulat, "Thermal-photovoltaic solar hybrid system for efficient solar energy conversion," Solar Energy, vol. 80, no. 2, pp. 170–176, 2006.
149. Y. V. Vorobiev, J. González-Hernández, and A. Kribus, "Analysis of potential conversion efficiency of a solar hybrid system with high-temperature stage," Journal of Solar Energy Engineering, vol. 128, no. 2, pp. 258–260, 2006.
150. S. Jiang, P. Hu, S. Mo, and Z. Chen, "Optical modeling for a two-stage parabolic trough concentrating photovoltaic/thermal system using spectral beam splitting technology," Solar Energy Materials and Solar Cells, vol. 94, no. 10, pp. 1686–1696, 2010.
151. L. T. Kostic, T. M. Pavlovic, and Z. T. Pavlovic, "Influence of reflectance from flat aluminum concentrators on energy efficiency of PV/Thermal collector," Applied Energy, vol. 87, no. 2, pp. 410–416, 2010.
152. Y. Xuan, X. Chen, and Y. Han, "Design and analysis of solar thermophotovoltaic systems," Renewable Energy, vol. 36, no. 1, pp. 374–387, 2011.
153. G. Kosmadakis, D. Manolakos, and G. Papadakis, "Simulation and economic analysis of a CPV/thermal system coupled with an organic Rankine cycle for increased power generation," Solar Energy, vol. 85, no. 2, pp. 308–324, 2011.
154. H. Huang, Y. Li, M. Wang et al., "Photovoltaic-thermal solar energy collectors based on optical tubes," Solar Energy, vol. 85, no. 3, pp. 450–454, 2011.
155. L. T. Yan, F. L. Wu, L. Peng et al., "Photoanode of dye-sensitized solar cells based on a ZnO/TiO2 composite film," International Journal of Photoenergy, vol. 2012, Article ID 613969, 4 pages, 2012. ·
156. S. Kumar and A. Tiwari, "An experimental study of hybrid photovoltaic thermal (PV/T)-active solar still," International Journal of Energy Research, vol. 32, no. 9, pp. 847–858, 2008.
157. S. Kumar and G. N. Tiwari, "Life cycle cost analysis of single slope hybrid (PV/T) active solar still," Applied Energy, vol. 86, no. 10, pp. 1995–2004, 2009.
158. S. Kumar and A. Tiwari, "Design, fabrication and performance of a hybrid photovoltaic/thermal (PV/T) active solar still," Energy Conversion and Management, vol. 51, no. 6, pp. 1219–1229, 2010.
159. M. K. Gaur and G. N. Tiwari, "Optimization of number of collectors for integrated PV/T hybrid active solar still," Applied Energy, vol. 87, no. 5, pp. 1763–1772, 2010.
160. G. N. Tiwari, S. Nayak, S. Dubey, S. C. Solanki, and R. D. Singh, "Performance analysis of a conventional PV/T mixed mode dryer under no load condition," International Journal of Energy Research, vol. 33, no. 10, pp. 919–930, 2009.
161. G. Mittelman, A. Kribus, and A. Dayan, "Solar cooling with concentrating photovoltaic/thermal (CPVT) systems," Energy Conversion and Management, vol. 48, no. 9, pp. 2481–2490, 2007.
162. G. Mittelman, A. Kribus, O. Mouchtar, and A. Dayan, "Water desalination with concentrating photovoltaic/thermal (CPVT) systems," Solar Energy, vol. 83, no. 8, pp. 1322–1334, 2009.
163. H. Zondag, J. Bystrom, and J. Hansen, "PV-thermal collectors going commercial," IEA SHC Task 35 paper, 2008.

164. J. Hansen, H. Sørensen, J. Byström, M. Collins, and B. Karlsson, "Market, modelling, testing and demonstration in the framewrk of IEA SHC Task 35 on PV/thermal solar systems," in Proceedings of the 22nd European Photovoltaic Solar Energy Conference and Exhibition, DE2-5, Milan, Italy, September 2007.

165. C. L. Shen and J. C. Su, "Grid-connection half-bridge PV inverter system for power flow controlling and active power filtering," International Journal of Photoenergy, vol. 2012, Article ID 760791, 8 pages, 2012. ·

166. D. Amorndechaphon, S. Premrudeepreechacharn, K. Higuchi, and X. Roboam, "Modified grid-connected CSI for hybrid PV/wind power generation system," International Journal of Photoenergy, vol. 2012, Article ID 381016, 12 pages, 2012. ·

167. M. J. Elswijk, H. A. Zondag, and W. G. J. van Helden, "Indoor test facility PVT-panels—a feasibility study," Energy Research Centre of the Netherlands, ECN-I-02-005, 2002.

PART II

SOLAR ENERGY AND CONSERVATION

CHAPTER 2

AN APPROACH TO ENHANCE THE CONSERVATION-COMPATIBILITY OF SOLAR ENERGY DEVELOPMENT

D. RICHARD CAMERON, BRIAN S. COHEN, AND SCOTT A. MORRISON

2.1 INTRODUCTION

Climate change poses one of the greatest threats to biodiversity [1], [2]. Many species will be challenged to adapt to the magnitude and pace of the change, especially those already compromised by habitat loss and degradation [3]. Conservation of biodiversity will rely on protecting and enhancing the resilience and permeability of landscapes, to increase the viability of native species and provide them access to conditions they will need to persist in the future [4]. Efforts to reduce greenhouse gas emissions will also provide benefits to natural systems by reducing the magnitude of climate change impacts to which they need to adapt. Indeed, development of utility-scale (>1 MW) renewable energy generation facili-

An Approach to Enhance the Conservation-Compatibility of Solar Energy Development. © *Cameron DR, Cohen BS, Morrison SA. PLoS ONE 7,6 (2012); doi:10.1371/journal.pone.0038437. Licensed under the Creative Commons Attribution 3.0 Unported License, http://creativecommons.org/licenses/by/3.0/.*

ties is a core element of a multi-faceted strategy to reduce emissions from the energy sector [5]. Yet, such facilities can have sizable footprints in terms of land area and water use [6], and so can threaten natural ecosystems directly through habitat loss and fragmentation, or indirectly through the displacement of other human land uses [7]. Therein lies a paradox of utility-scale renewable energy development: it may be necessary to reduce climate change impacts and help protect biodiversity worldwide in the future; but if not carefully planned, it could come at the expense of the viability of local species today or constrain their ability to adapt to future conditions by destroying, or creating dispersal barriers to, areas they will need in the future.

The current pace and scale of efforts to develop renewable energy sources can make it more difficult to avoid adverse ecological impacts, especially given the lack of scientific studies regarding those impacts [8]. Yet, if emissions levels are to be maintained below what some describe as "dangerous" for both natural and human systems [9], [10], conversion to renewable sources of energy needs to be rapid worldwide [11]. Interest in energy security and economic stimulus further fuels demand for renewable energy development in the United States. Utility-scale development has become a government priority at the national and subnational level, with regulatory and financial incentives to further it (examples include the National Energy Policy Act of 2005, American Reinvestment & Recovery Act of 2008) including $5.3 B in loan guarantees for three projects in California [12]. This has resulted in a boom market for renewable energy in the western United States that has overwhelmed state and federal environmental regulatory processes and permitting agencies. For example, as of November 2010, there were 22 applications to develop solar facilities on Bureau of Land Management (BLM) lands in the California deserts alone, with a cumulative footprint of nearly 78,000 ha [13].

Regulatory complexity compounds the political and market pressures. Authority for permitting new renewable energy facilities is dispersed across multiple jurisdictions depending on the technology, the size of the facility being proposed, and whether the proposed location is on public or privately-owned land. A variety of undesired consequences may result from this high political pressure and complexity, including protracted and controversial approval processes, unexpectedly high compensatory mitigation costs, and approval of projects prior to a full understanding of their cumulative environmental impact.

Decision-support tools are needed to efficiently guide projects toward areas that are commercially attractive for development, and away from areas important for biodiversity conservation and other resources. Using such tools in the early phase of project scoping would allow developers to select areas where they will be less likely to encounter environmental obstacles in the permitting process. These "low-conflict" locations could be prioritized for field investigations and possibly be eligible for expedited permitting or other incentives to promote projects on appropriate lands. Conservationists also benefit from early identification of areas with

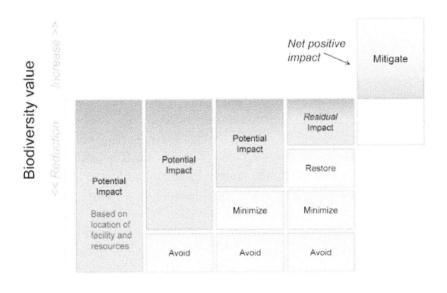

FIGURE 1: The mitigation hierarchy. Under this schema, developers advancing a project choose locations for their project that avoid environmental impacts. If impacts cannot be completely avoided, they then take steps to minimize impacts. Once impacts are minimized to the extent possible, restoration opportunities are pursued. Residual impacts not addressed by the previous steps are then offset through compensatory mitigation, using ratios that result in a net positive impact on biodiversity. Adapted from Convention on Biological Diversity 2008 [54].

minimal conservation value as it might expedite the attainment of climate benefits and reduce the risk of their being perceived as obstructionist.

Avoiding impacts through the selection of appropriate development locations and compensating for any residual impacts are core components of the "mitigation hierarchy", a planning approach most commonly used to avoid impacts to wetlands [14], [15] (Figure 1). Adherence to this approach can help reduce adverse impacts of development, by defining resources and areas to be avoided, and outlining steps to minimize, restore, or offset unavoidable impacts. The principles of the mitigation hierarchy can be applied at a landscape scale through spatial analyses that map constraints and opportunities for both development and conservation [16]–[18]. Finding areas that are both suitable for renewable energy development and of relatively low biodiversity conservation value represents a possible "win-win" for two otherwise potentially conflicting objectives [19]. When complete avoidance of impacts is not possible, this approach can improve the conservation return of investments in compensatory mitigation, by directing it to places and efforts that also advance regional conservation goals [16], [20], [21].

Here, we illustrate how the mitigation hierarchy can be applied to characterize the degree of alignment between biodiversity conservation and electricity generation from utility-scale solar facilities. Our study focuses on the Mojave Desert, as it is the focus of intense development pressure: it offers large expanses of public lands with exceptional solar energy resources in close proximity to highly populated regions with strong markets for renewable energy. We integrate conservation values and presumed development feasibility across the desert, and illustrate how compensatory mitigation can contribute to regional conservation goals. We propose that this regional application of the mitigation hierarchy can lead to both more efficient development of renewable energy and better conservation outcomes in the Mojave Desert, and that this approach can serve as a model for resolving such conflicts more generally.

2.1 STUDY AREA

The Mojave Desert Ecoregion encompasses 13,013,000 ha, across four southwestern states: California (contains 56% of the ecoregion), Nevada

(31%), Arizona (11%) and Utah (2%). The ecoregion is notable for its biodiversity as well as for its wilderness values and associated economic benefits [22]. There are over 400 vertebrate species that inhabit the ecoregion, with extremely high endemism especially in wetland areas, such as Ash Meadows National Wildlife Refuge, Nevada where there are 24 endemic plants and animals [23], [24]. Plant diversity in shrub communities is among the highest in North America, with potential species diversity in these communities as high as 70 species per hectare in the eastern Mojave [25]. Currently 29 species and subspecies in the Mojave Desert are listed as threatened or endangered under the federal Endangered Species Act [23]. The region has extensive public and military lands (collectively covering over 85% of the ecoregion), with 53% of the ecoregion designated for wilderness and for species habitat—such as critical habitat for the federally threatened desert tortoise (*Gopherus agassizii*).

The biodiversity input into this analysis is a characterization of conservation value across the Mojave Desert Ecoregion, from Randall et al.'s (2010) Mojave Ecoregional Assessment (hereafter, the Assessment) [26]. The Assessment analyzed a broad set of conservation elements, or "targets" (44 vegetation communities and 521 plant and animal taxa) and used the conservation planning software Marxan [27] to generate alternative configurations of areas to meet conservation objectives. By integrating Marxan output of priority areas, aerial photo interpretation (to assess degree of anthropogenic ground disturbance), and principles of conservation reserve design, Randall et al. classified the land into categories of high (i.e., Ecologically Core, Ecologically Intact) and low (i.e., Moderately Degraded, Highly Converted) conservation value (Figure 2). Here, we used the latter category to represent areas of lower conservation value. We note that the approach we present is flexible, and could accommodate other conservation assessments as the biodiversity input. For example, other prioritization analyses exist for individual species in the ecoregion (such as federal endangered species critical habitat units) or as habitat conservation plans for portions of the ecoregion [28]–[29]. We selected the Randall et al. 2010 conservation value assessment because it is the most recent and consistent characterization of the distribution of biodiversity and land use impacts across the whole of the ecoregion.

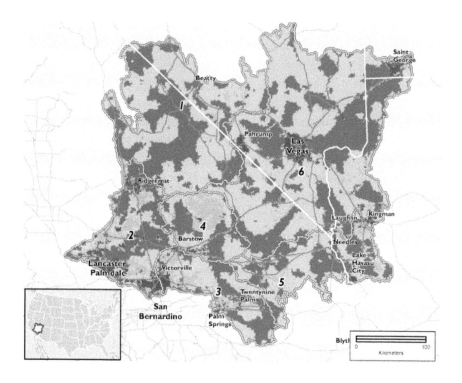

FIGURE 2: Conservation value in the Mojave Desert Ecoregion. The conservation values categories are depicted on the map as follows: dark gray areas are Ecologically Core, light gray are Ecologically Intact, orange are Moderately Degraded, and the darkest patches are Highly Converted (adapted from Randall et al. 2010). Subregions of the Mojave Desert are shown in the outline; labels indicate the 1. Northern, 2. Western, 3. South-central, 4. Central, 5. Southeastern, and 6. Eastern subregions. Urbanized land is grey and highways are in grey lines. The location of the ecoregion in the coterminous United States is shown in the inset map.

The Mojave Desert is also renowned for its extraordinary solar resources. An analysis of the solar energy production potential of the southwestern United States suggests that the region could supply 50% of the country's electricity demand if fully developed [30], [31]. One of the largest collections of solar electricity facilities in the world, the Solar Energy Generating Systems (SEGS) is installed in the Mojave Desert, totaling 354 MW of installed capacity.

Certain attributes of the desert ecosystem warrant special attention in planning for industrial land uses such as energy facilities. The low productivity of the desert leads to a slow pace of soil development, plant growth, and ecological succession, and that renders it slow to recover from disturbances [32]. This limits the application of the mitigation hierarchy, in that restoration of disturbed areas is often infeasible in ecological timeframes. While restoration is a critical step for reducing impacts from infrastructure development in many ecosystems, the challenges of successful restoration in desert systems increases the importance of avoidance and minimization strategies. Mechanical disturbance of soil crusts leads to erosion and heightened susceptibility to invasion by non-native grasses and forbs [33]. Those, in turn, can result in altered fire regimes, and effectively irreversible type conversion of habitats [34]. Disturbing desert soil may also limit the degree to which it acts as a carbon sink, an ecological process that is poorly studied and the magnitude of which has only recently been characterized [35]. Solar facilities also consume water in their installation, operation, and or maintenance. Water is very limiting in the desert, with many species dependent upon either the rare surface expressions of water or the vegetation communities that draw upon subsurface flows. Although relationships between surface and ground water, as well as ground water flows and recharge rates are poorly understood, it is generally accepted that these resources are over-allocated [36]. While a full consideration of the ecological values of desert ecosystems is beyond the scope of this study (see Lovich and Ennen 2011), the integrity of soils and the scarcity of water are two key ecological attributes for planning, and potential constraints on the ability to align solar energy development and biodiversity conservation.

2.2 RESULTS

2.2.1 REGIONAL OPPORTUNITIES TO ALIGN ENERGY AND CONSERVATION GOALS

We found large areas of the Mojave Desert that are potentially suitable for the development of solar facilities that are ecologically degraded with

FIGURE 3: Conservation values in potentially suitable lands for solar development below 5% slope angle. Urban areas, water bodies, and lands outside of private or BLM multiple use ownerships, and areas above 5% slope were removed. Conservation value colors are the same as Figure 2. Lands in darkest gray are classified as lower conservation value lands for which energy production estimates are provided in the results.

lower regional conservation value (Figure 3). The amount of lower conservation value land that meets the development suitability criteria ranges from nearly 200,000 ha (<1% land surface slope angle) to over 740,000 ha (<5% slope) (Table 1). The level of potential compatibility between development and conservation is much greater if land with higher slope can be utilized, with nearly four times more lower conservation value land at the 5% cutoff compared to the 1%.

TABLE 1: Area (ha) of land by land ownership, conservation value, and percentage slope angle.

Owner	Ecologically Core	Ecologically Intact	Moderately Degraded	Highly Converted	Total
Bureau of Land Management (BLM)					
<5% slope	389, 458	828, 371	190,244	21,669	1,429,742
<3% slope	240,370	491,398	130,530	17,762	880,061
<1% slope	73,736	99,196	31,785	10,570	215,288
Private Land					
<5% slope	159,693	221,835	400,264	128,552	910,315
<3% slope	128, 260	168,127	326,898	111,955	735,239
<1% slope	49, 045	34,811	89,886	58,687	232,428

Areas with lower than 7 kwh/m²/day direct normal irradiance (DNI) were excluded from the analysis, as were legally and administratively protected areas, urban areas, and perennial water bodies. BLM land includes only undesignated land eligible for potential siting. Higher percentage slope categories are inclusive of the lower. Conservation value categories from Randall et al. 2010.

Privately-owned parcels provide considerably more opportunity to develop on land with lower conservation value than do public lands (Figure 4, Table 1). The combined area of lower conservation value private land is 3.5 times (<1% slope) to 2.5 times (<5% slope) the area of those categories on suitable BLM land across the ecoregion. The higher degradation on private land is primarily due to agricultural land use and low density development in parts of the western Mojave in California and in the Arizona portion of the ecoregion. However, unlike BLM-managed lands, private lands are often parcelized and divided into many ownerships. In California, private lands that meet suitability criteria, are less than 5% slope and are in the lower conservation value categories, the average parcel size is 2.4 ha, with a median of 1 ha (Figure 5).

While most of the degraded areas potentially suitable for development are found on private land, BLM land also provides large areas of potential opportunity for development, with over 210,000 ha of lower conservation value land less than 5% slope across the ecoregion (Table 1, Figure 4). About 90% of those lands are available for solar use since approximately

10% (21,522 ha) are within designated off highway vehicle (OHV) open areas and thus likely to be off limits to and inappropriate for development.

2.2.2 ECOREGIONAL IMPACTS

If the full extent of areas without protective designation (i.e., BLM multiple use and private lands) that are potentially suitable for solar facilities were to be opened and used for solar development, large areas of Ecologi-

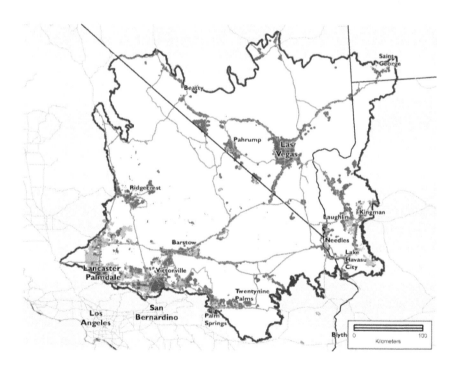

FIGURE 4: Land ownership in potentially suitable lands below 5% slope with Moderately Degraded and Highly Converted conservation value. Lighter gray areas are private lands and dark gray areas are BLM land without designation. Areas outlined in medium gray are designated open off-highway vehicle areas on BLM land in California, accounting for 10% of the 211,000 ha in lower conservation value on BLM land and would not be suitable for development. Conservation values adapted from Randall et al. 2010.

cally Core and Intact (hereafter, "higher conservation value") lands would be lost, ranging from over 250,000 ha (<1%) to 1.6 million ha (<5%) (Table 1). This extent of loss would greatly reduce the ability to meet ecoregional conservation goals (per Randall et al. 2010) for many biodiversity targets, especially if higher slopes are eligible for development (Figure 6). Some targets would face extensive loss relative to the current distribution, such as mesquite upland scrub, greasewood flats, blackbrush shrubland, and mixed salt desert scrub [37] (Figure 6). The extent of desert tortoise suitable habitat outside tortoise conservation areas in higher conservation value lands that would be lost varies considerably based on slope angle, from 90,103 ha (<1%) to over 1 M ha (<5%). The location of many of the areas at risk are in flat valleys which often connect existing conservation lands for wide-ranging species like desert bighorn sheep (*Ovis canadensis nelsoni*) [38].

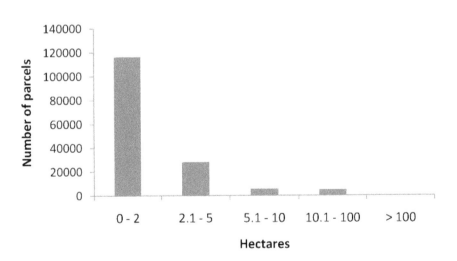

FIGURE 5: Parcel size class distribution within private lands of California that are of lower conservation value. These are only within areas that are potentially suitable for solar development below 5% slope. The presence of high rates of parcelization on private land acts as a disincentive to site large solar projects in more degraded areas.

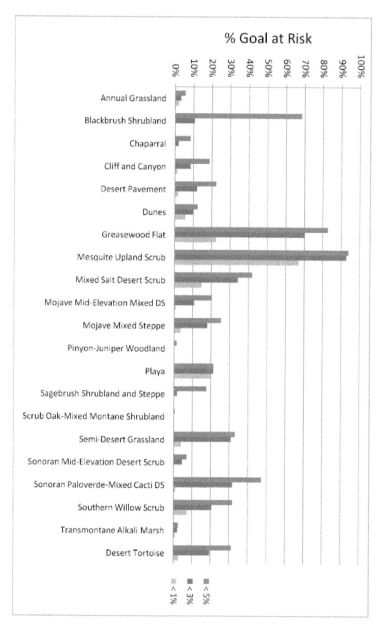

FIGURE 6: Percent of representation goals that would not be attainable if all areas potentially suitable for solar development were to be developed. The goals refer to a hypothesized amount of each habitat that needs to be managed for conservation to meet long-term viability needs for representative biodiversity of the ecoregion. Goals are based on Randall et al. 2010.

In the California and Nevada portion of the Mojave, there are over 220,000 ha of solar facilities proposed as Right of Way (ROW) applications on BLM lands, including nearly 130,000 ha of Ecologically Core and Intact habitats (Table 2). The vast majority of this area—over 116,000 ha—is occupied by the ecoregion's most widespread community, creosotebush-white bursage desert scrub (*Larrea tridentata, Ambrosia dumosa*). The second most extensive impact would be to Mojave mid-elevation mixed desert scrub [37] (Table 2). The desert tortoise is wide-ranging across the study area, and would directly lose 103,509 ha of Ecologically Core and Intact suitable habitat if the footprints of all current proposals on BLM lands are developed.

TABLE 2: The extent of ecological system targets that occur within BLM Right of Way applications in California or Nevada that also occur within Ecologically Core or Ecologically Intact conservation value categories (from Randall et al. 2010).

Conservation Target	Area Potentially Impacted (Ha)
Creosotebush-White Bursage Desert Scrub	116,640
Mojave Mid-Elevation Mixed Desert Scrub	7,125
Southern Willow Scrub	1,145
Mixed Salt Desert Scrub	1,082
Cliff and Canyon	1,075
Playa	699
Desert Pavement	578
Dunes	229
Greasewood Flat	165
Chaparral	110

2.2.3 SUPPLY RELATIVE TO RENEWABLE ENERGY GOALS

California's 2020 Renewables Portfolio Standard (RPS) goal can be fully met without developing within the Ecologically Core or Intact lands in the ecoregion. The lower conservation value land with slopes of less than 1% (190,928 ha) could supply 107 TWh of electricity, or 180% of the renewable energy that it is estimated will be needed to meet the California RPS by the Renewable Energy Transmission Initiative (RETI) [39]. Below 3% slope, there are 587,145 ha of land, with the potential to generate 555% of the energy required, while the lower conservation value lands below 5% cutoff (740,699 ha) could supply 700% of the energy required.

2.2.4 MITIGATION SCENARIOS

We calculated a total footprint of 31,994 ha for proposed solar energy generation facilities under verified Right of Way applications on BLM lands and on private lands of the western, central and south-central subregions of the ecoregion. Meeting compensatory mitigation needs for these proposed projects would contribute more to regional conservation goals if mitigation is not restricted to private lands. For example, if we use the "future" mitigation ratio and restrict mitigation investment to private lands, there will not be enough higher conservation value private land in the central Mojave subregion to offset impacts for five conservation targets, including the desert tortoise, which falls short of the mitigation need by 38% (23,104 ha) (Figure 7). In contrast, if public lands are also eligible for investment, mitigation requirements under the future ratio could be met for all but two targets (playa is short by 601 ha and desert pavement is short by 30 ha) (Figure 7). Moreover, in the private land only scenario, lands selected for mitigation at both ratio levels are more fragmented than the mixed ownership scenario (as reflected in higher edge length of the full selected network, 15% higher for current ratios and 52% for future ratios). The areas selected in the private land only, current scenario are slightly more degraded (11%, as indicated by the average Marxan "cost" per selected assessment unit) than the mixed ownership solution (Table 3). This difference in degradation jumps to 60% using the future ratios, which is largely due to Marxan seeking to meet the mitigation goals for tortoise, by having to include areas that may be relatively more impacted.

The ideal arrangement of places for mitigation differs depending on what lands are available. The percentage overlap of the mitigation solutions for the mixed ownership and the private land only scenarios is low: the Jaccard similarity index [40] was 0.29 for the current mitigation ratio and 0.42 for the future ratio (Figure 7). A similar comparison of total area

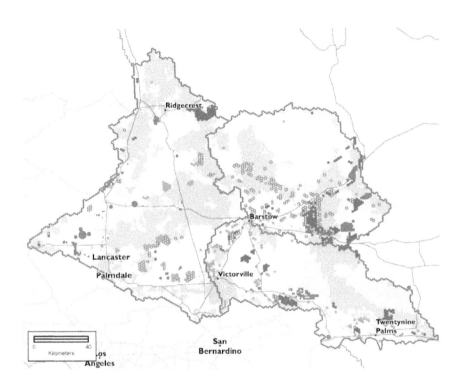

FIGURE 7: Scenarios of suitable mitigation areas using the future ratios. This map shows the private land-only (medium gray) and the mixed ownership (dark gray) scenarios, with planning units that are shared in both scenarios (with outline). The private land-only solution is more dispersed and was not able to offset impacts for five targets in a subregion (grey outlines, labeled in Figure 2), most notably a deficit of over 23,000 hectares of suitable desert tortoise habitat in the Central Mojave subregion, north and east of Barstow, CA. Urbanized areas are shown in lighter grey. The extent of Ecologically Core (darker green) and Ecologically Intact (lightest gray) is shown for reference (adapted from Randall et al. 2010).

needed for both ownership scenarios could not be performed for the future ratio solutions because mitigation goals could not be met in the private land only, future scenario (Table 3).

TABLE 3: Performance of compensatory mitigation scenarios.

Mitigation Ratios	Eligible Land for Mitigation	Assessment Unit Cost	# of Assessment Units	Boundary Length (m)	Goals Met for All Targets?
Current	Private Core and Intact	158,999	254	1,087,545	No (1 not met)
Current	Private or BLM undesignated Core and Intact	141,084	251	717,062	No (1 not met)
Future	Private Core and Intact	447,275	457	1,862,370	No (5 not met)
Future	Private or BLM undesignated Core and Intact	324,674	531	1,617,374	No (2 not met)

Assessment unit costs are the sum of the "cost" values, a unitless index used in Marxan as a proxy for anthropogenic disturbance. The number of assessment units is the number selected in the most efficient scenario of 100 model runs. Boundary length is the total edge length of the selected assessment units and is a proxy for the dispersion of the selected network of areas. Goal attainment refers to whether the mitigation goals for the targets are met in the given scenario. See Supporting Information S2 for full description of Marxan settings.

2.3 DISCUSSION

We found considerable opportunity for alignment of biodiversity conservation and solar energy development objectives in the Mojave Desert. Assessed at the moderate 3% slope cutoff, over 580,000 ha of lands with lower conservation value yet presumably suitable for solar energy development currently exist across the desert, an amount that could supply over five times the energy needed to meet the projected 2020 California 33% RPS goal. Steering development to areas of lower conservation value could help reduce adverse impacts to desert ecosystems, specifically areas that are more intact and those that contain sensitive resources. Avoiding those areas will likely improve the adaptive capacity of desert species in the face of climate change and provide greater ecological resilience in the future. Prioritizing development in lower conservation value lands reduces the prospect of conflict over ecological impacts that can add cost, delay, and controversy to projects.

One striking finding from this study is the relationship between land ownership, conservation value, and "attractiveness for development." From a conservation perspective, most of the areas that appear better suited for development are privately held, but they are often comprised of many parcels that would need to be consolidated to achieve a minimum area sufficient to support a project. From a development perspective, that parcelization creates a disincentive, especially if an alternative exists to have a more streamlined process working elsewhere with one land owner, e.g., BLM. Thus, one strategy to enhance protection of the conservation values of the Mojave Desert would be to develop policies that incentivize development on degraded private lands. We note that brownfields and areas formerly in agricultural production, but retired due to salinity, water limitations, economic considerations, or other contamination problems may present ideal locations for solar development, especially for technologies that use less groundwater than the former land use.

The approach we present can also help direct compensatory mitigation investments. By accounting for the direct impacts of a given set of proposed projects and the distribution of lands with higher conservation value, we illustrate how one can generate a portfolio of candidate areas for compensatory mitigation that meet mitigation obligations while contributing to regional conservation goals. Of course, further field assessment is required to ensure that candidate sites generated from this type of analysis are indeed suitable as mitigation. This approach can be generalized to other land uses, geographies, covered resources, and mitigation ratios and actions, and explored as a site-selection problem to optimize various social and ecological goals.

Our analysis of land ownership and conservation value also revealed a conundrum for mitigation. While the higher degradation of private lands provides opportunities to avoid or minimize adverse ecological impacts when siting projects, it also poses problems if compensatory mitigation can only be conducted on private lands. The limited supply of private lands with higher conservation values could in turn limit the amount of energy development for which impacts can be offset. We note, however, that there may be considerable opportunity to use mitigation funds to enhance the conservation management of existing public lands in the desert, through such actions as eradicating invasive species, increasing enforcement of off-highway vehicle closures, or installing tortoise exclusion fencing along roads. The desert tortoise recovery plan [41], for example, recommends numerous management actions to enhance species viability, many of which go unimplemented due to insufficient funding [42]. We emphasize that any investment of mitigation resources applied to public lands would need to result in enduring conservation outcomes and add to the current level of management activities rather than replace existing resources and agency obligations. One way to track and better ensure that investments result in enduring conservation is to change the designation of lands serving as mitigation from one that allows multiple uses to one that gives primacy to the conservation use. Ensuring additionality of mitigation-related enhanced management funding would likely involve contractual obligations and require special enforcement mechanisms within agency budgeting processes.

We underscore the importance of accounting for cumulative impacts in siting and mitigation decisions, especially in light of the increased stress that climate change will exert on desert ecosystems. The impacts of projects should not only be evaluated comprehensively regarding ecological impacts, but also examined cumulatively in the context of all of the major stressors in the desert (including but not limited to the other proposed energy projects). Because of the large area potentially impacted by long-term solar energy development (as illustrated in Figure 6), and the lack of related impact studies, a framework is needed in the near term to guide decision-making to help reduce the risk of inadvertently crossing thresholds of ecological viability [8]. The approach presented here, essentially an application of the precautionary principle, can provide that initial guidance: develop first in the least conflict areas and protect the consensus conservation areas; meanwhile, improve knowledge regarding the areas in between, so that siting and mitigation decisions in the future can be better informed as to their environmental trade-off.

Limitations of this analysis are mostly related to data quality and resolution. We underscore that this study cannot substitute for site-level assessment, or more detailed assessments of sensitive and rare species' conservation needs (e.g., HCPs [Habitat Conservation Plans], NCCPs [the state of California's Natural Communities Conservation Plans], endangered and threatened species recovery plans). Moreover, the map of the relative conservation value should not be construed as a development and conservation blueprint, per se. Randall et al. (2010) caution that because important occurrences, ecological processes or habitats of targets may occur within all of the conservation value categories, even the Highly Converted category, site-level assessment is needed to confirm suitability for development, and guide project siting, design, and mitigation. The Assessment is best used to provide general guidance to planners and industry seeking to assess the relative likelihood of environmental constraints across a broad area, in an attempt to minimize adverse permitting problems. As suitable information becomes available, the approach we present here can be implemented at a finer spatial scale for a portion of the ecoregion.

An additional limitation of our analysis is that it does not explicitly account for some key factors that influence the economic feasibility of project development. Geographic factors may affect the economic profitability of a site, such as local influences on solar radiation or the costs of ongoing maintenance to minimize damage from airborne sand. One notable factor that was beyond the scope of our study pertains to transmission. Proximity to transmission corridors that have additional capacity is an important consideration in siting new generation facilities. The relationship between transmission and generation will be important to incorporate into future refinements of this analysis utilizing the expertise of the solar industry, especially where new transmission is required to service proposed facilities. Those additional impacts should be incorporated into the overall application of the mitigation hierarchy.

In sum, we demonstrate how solar energy production goals in the Mojave Desert can be met with less adverse effect on biodiversity. The systematic approach presented here for proactively balancing solar energy production with biodiversity protection better accounts for, and so can help reduce, trade-offs. Importantly, it can also provide greater assurances to agencies, developers and conservationists that their respective goals are being met. Integrating this sort of analysis with dynamic information systems for species distributions, ecological condition and conservation investments, can help agencies and stakeholders adaptively apply the mitigation hierarchy with increasing effectiveness. This example of multiobjective planning can also be expanded and tailored to other technologies and geographies, e.g., wave energy and marine protected areas. We caution, however, that if such planning does not incorporate and accommodate all major interests and stakeholders, it may lead to displacement of one user by another, and exacerbate rather than resolve conflict. For example, our analysis did not incorporate some significant desert values, such as cultural values, recreational uses, military training, and scenic values. Accounting for this array of interests will be essential for developing the long-term conservation plan for the Mojave.

Numerous conservation and energy development planning efforts are currently underway that will affect the Mojave Desert (e.g., BLM's Solar Energy Development Programmatic Environmental Impact Statement). The State of California is currently developing an NCCP for the state's deserts that, like this analysis, will take into account not just those species currently listed but the full array of natural communities of the California deserts. We are hopeful that the resulting NCCP will identify areas preferred for development and conservation, and institutionalize effective regulatory mechanisms and market-based incentives to implement that plan. Ideally, those mechanisms will help ensure that siting and mitigation occur in the places most appropriate for effecting desert conservation–regardless of the underlying ownership. In the interim, we propose that a precautionary approach like that presented here could guide conservation-compatible renewable energy development in the desert.

2.4 MATERIALS AND METHODS

2.4.1 SOLAR ENERGY DEVELOPMENT POTENTIAL

We estimated solar energy potential across the Mojave Desert using the direct normal irradiance (DNI) data at 10 km resolution developed by National Renewable Energy Laboratory (NREL) and SUNY-Albany [43].

The DNI is the variable commonly used to assess the potential for concentrating solar power (CSP) installations, but is strongly correlated with solar insolation values used to plan solar photovoltaic (PV) facilities.

Development feasibility was characterized based on land ownership and management, current land use, and land surface percent slope angle, as well as solar insolation. We filtered the DNI data to include only those lands with excellent solar resource potential (annual average value of at least 7 kWh/m²/day) and slope angles that bracket the maximum slope that is considered to be developable for solar energy based on current technologies (less than 1%, 3% (inclusive), and 5% (inclusive)). We calculated the slope using elevation data from the Shuttle Radar Topography Mission (SRTM) resampled from 30 meters to 90 meters resolution, and smoothed using an averaging filter by a 3 × 3 window to remove anomalies in the data [44]. To remove patches of land not large enough for utility-scale solar projects, we applied a minimum mapping unit of 100 hectares and merged all polygons below this cutoff with adjacent polygons using the ARCGIS Eliminate tool [45].

To ensure that areas already developed with residential, industrial or commercial uses were not included as potentially suitable, we created a composite "developed" land layer. For Utah, Arizona and Nevada we used data from the Southwest ReGap program [46] to represent developed land use. For California, we extracted the "urban" category from the Multi-source Land Cover data [47] to represent the footprint of areas to exclude. To minimize adjacency to urban areas, we smoothed the developed land composite using an averaging filter by a 3×3 window and removed all areas greater than 10% urbanized after smoothing. Perennial water bodies and areas that have a legal or administrative status that prevents energy development were also removed from the suitable land base. We removed the categories of land that were identified as consensus exclusion areas in California's Renewable Energy Transmission Initiative [48] (see Supporting Information S1 for a list of these categories). We also excluded the desert tortoise conservation areas as defined by the U.S. Fish and Wildlife Service, which include areas designated as critical habitat for the desert tortoise [49]. Mohave ground squirrel (*Spermophilus mohavensis*)) conservation areas [50] were also removed because they have been proposed for exclusion by the BLM. Management status data on the location of public and private land and the relative level of conservation management were from the U.S. Geological Survey Protected Areas Data version 1.1 [51].

2.4.2 PROGRESS TOWARD CALIFORNIA RENEWABLE ENERGY GOALS

A key driver of demand for renewable energy in the Mojave Desert is the California Renewables Portfolio Standard (RPS), which mandates that investor- and publicly-owned utilities acquire 33% of their energy from renewable sources by 2020. The net amount of renewable energy that needs to come online to meet the 2020 goal will change over time and requires assumptions about the lifespan of current and future projects. We used an estimate from the California Renewable Energy Transmission Initiative [39] of 59.7 TWh which is higher than more recent estimates [52]. We calculated the potential energy generation based on the land area that is developable based on the solar insolation, slope, and land use and management filters described above, and conservation value (per Randall et al. 2010) for the whole ecoregion. We used this potential energy generation to estimate the proportion of the remaining California's RPS goal (net short) that could be met in the Moderately Degraded and Highly Converted (hereafter, "lower conservation value") lands in the ecoregion. We considered the California RPS as a realistic energy goal for this analysis, and we assumed that land in other states can have projects to contribute to the California RPS goal given the close proximity of many of the areas to California. To convert land area to energy output, we used the mid-point land area to energy estimate for solar thermal provided in MacDonald et al. (2009) of 3.8 ha/mw and assumed a 25% capacity factor [7].

2.4.3 DEVELOPMENT IMPACTS AND MITIGATION OPPORTUNITIES

We analyzed opportunities to offset projected impacts from BLM and private land solar projects by developing mitigation scenarios that differed in 1) the type of land ownership allowed to serve as mitigation, and 2) the mitigation offset ratio. The extent of this analysis included three sub-regions used in the Assessment: the Western, Central, and South Central Mojave Desert (Figure 2). We used only the northern portion of the South Central subregion (dividing it based on the ecological subsection boundary [53]) because the southern portion is covered by Joshua Tree National Park and an adjacent Area of Critical Environmental Concern (ACEC), which are land designations that do not allow for development.

To estimate subregional impacts, we used the mapped or estimated footprints of proposed solar projects on private lands in Kern, San Bernardino, and Los Angeles counties within the California Mojave ecoregional boundary and the verified ROW applications for BLM lands in California [13]. For the BLM projects, we used the California verified Right of Way solar projects from a data download from November 8, 2010. For the private land projects, we used maps or available GIS data from Kern, Los Angeles and San Bernardino counties. Specifically, for Kern County projects was a spreadsheet and digital map showing the location of the facilities, acquired from the county and dated September 9, 2010. The facilities were digitized based on this map and a point GIS file was created. The area of the facility was used from the spreadsheet to buffer the point to a circle with an area the exact same size as the listed size in the table. The source for San Bernardino County projects was from April 2010 and included two pre-application projects. These were digitized based on the locations and information in a digital map acquired from the county. We mapped the projects as precisely as possible to get the approximate acreage and location based on the information available, though we were not able to map projects more accurately than the parcel boundary. For Los Angeles County, projects were mapped based on available assessor parcel numbers and parcel data acquired in December 2010 from the county. The three county data layers and the BLM ROW layer were merged into one file within the extent of the subregional area. Each project was assigned to a subregion with no projects straddling subregions. We could not identify a data source for Inyo County in the western subregion.

To estimate potential ecoregional impacts from ROW applications, we included both California and Nevada applications. We assume that the whole area within the ROW would be impacted by the proposed projects, even though in many cases the area of the ROW application exceeds the actual development footprint. We caution that these footprints represent only the direct impacts associated with the projects, not indirect effects. It is also likely that not all of these applications will be developed. However, the purpose of this portion of the study is to characterize the magnitude of the impact of solar development based on a proposed set of projects and resultant mitigation it will require in one portion of the Mojave Desert.

To derive the amount of mitigation needed for species and vegetation system targets, we calculated the extent for each vegetation type and habitat for two species of conservation interest (desert tortoise [49], Mojave ground squirrel [50]) within the ROW applications and private land projects in the subregional study area. The calculated impacts for these 45 projects were used to identify potential areas to meet compensatory mitigation needs in the most efficient configuration (based on total area, length of outer boundary of selected hexagons, and conservation suitability described below) while contributing to regional conservation goals. We used the same tool for the mitigation scenarios that was used in the Assessment, Marxan (v. 1.8.10), to identify areas that can meet mitigation needs. We ensured that potential mitigation areas would contribute to conservation goals by allowing Marxan to select only Ecologically Core or Intact areas from the Assessment, without an existing protective designation, such as Federal Wilderness areas or Areas of Critical Environmental Concern. To ensure that the mitigation areas would be ecologically similar to the impacted resources, we required the offsetting to be within the same subregion as the impact. Additional parameters and goal amounts used for Marxan scenarios are shown in Supporting Information S2.

To assess mitigation needs, we used two sets of mitigation to impact ratios. The first set was intended to mitigate for the impacts of existing proposed projects (hereafter "current"). Current ratios were based on available guidance in existing regulations and recovery plans, although we included all target ecological systems, not just those for which mitigation is required under existing laws and regulations. The second set of ratios was intended to be a proxy for potential future build out of solar projects (hereafter "future"). "Future" ratios were defined as double the "current" ratios (Table 4). This simple approach to forecasting mitigation needs can be used to design programmatic investments, such as advance mitigation.

To assess the influence of land ownership on the availability of mitigation options, we ran scenarios with two alternatives: only using private land as suitable sites (hereafter "private land only") and using BLM multiple use land as well as private land as options (hereafter "mixed ownership"). To ensure that the mitigation areas selected had relatively minimal degradation, we used an index of anthropogenic disturbance (road density, urban and agricultural land) adapted from Randall et al (2010) to define conservation suitability as the "cost" layer input for Marxan. The details of this layer and the input data are shown in Supporting Information S3. Using this cost layer in the Marxan mitigation scenarios provided a basis for comparison of the relative habitat quality available using the two sets of allowable land ownerships for mitigation.

For desert tortoise habitat distribution, we used the output of the habitat model developed by Nussear et al. (2009) and selected the top four scores (>0.6) of the classified output as a conservative representation of higher quality habitat [49]. For Mohave ground squirrel, we used the boundaries of the conservation areas as designated by the BLM in California [50].

TABLE 4: Compensation ratios for current and future mitigation scenarios.

Target	Current Ratio	Future Ratio
Species	3:1	6:1
Vegetation Systems	2:1	4:1
Unvegetated Systems	1:1	2:1

Mitigation ratios represent the proportional offset needed per unit of impact. Current Ratio refers to a hypothetical degree of offset to compensate for impacts to the target species or system based on a set of proposed projects. Future Ratio refers to a potential amount of mitigation that might be needed based on future build out of solar projects. Unvegetated systems include dunes, cliff and canyon, desert pavement, and playas.

REFERENCES

1. Thomas CD, Cameron A, Green RE, Bakkenes M, Beaumont LJ (2004) Extinction risk from climate change. Nature 427: 145.148

2. Lovejoy TE, Hannah L, editors (2005) Climate change and biodiversity. New Haven, CT: Yale University.

3. Dawson TP, Jackson ST, House JI, Prentice IC, Mace GM (2011) Beyond Predictions: Biodiversity Conservation in a Changing Climate. Science 332: 53.58

4. Heller NE, Zavaleta ES (2009) Biodiversity management in the face of climate change: A review of 22 years of recommendations. Biological Conservation 142: 14.32

5. Edenhofer O, Pichs-Madruga R, Sokona Y, Seyboth K, Matschoss P (2011) IPCC Special Report on Renewable Energy Sources and Climate Change Mitigation. Cambridge, United Kingdom and New York, NY, USA: Cambridge University Press.

6. Tsoutsos T, Frantzeskaki N, Gekas V (2005) Environmental impacts from the solar energy technologies. Energy Policy 33: 289.296

7. McDonald RI, Fargione J, Kiesecker J, Miller WM, Powell J (2009) Energy Sprawl or Energy Efficiency: Climate Policy Impacts on Natural Habitat for the United States of America. PLoS ONE 4: e6802.

8. Lovich JE, Ennen JR (2011) Wildlife Conservation and Solar Energy Development in the Desert Southwest, United States. BioScience 61: 982.992

9. Mastrandrea MD, Schneider SH (2004) Probabilistic Integrated Assessment of "Dangerous" Climate Change. Science 304: 571.575

10. Schneider SH (2001) What is 'dangerous' climate change? Nature 411: 17.19

11. Pacala S, Socolow R (2004) Stabilization Wedges: Solving the Climate Problem for the Next 50 Years with Current Technologies. Science 305: 968.972

12. DOE US (2011) Loan Programs Office: Projects website.

13. BLM (2010) Right of Way Application database.

14. CEQ (2000) National Environmental Policy Act (NEPA) Regulations.

15. Kiesecker JM, Copeland H, Pocewicz A, McKenney B (2010) Development by design: blending landscape-level planning with the mitigation hierarchy. Frontiers in Ecology and the Environment 8: 261.266

16. McKenney B, Kiesecker J (2010) Policy Development for Biodiversity Offsets: A Review of Offset Frameworks. Environmental Management 45: 165.176

17. Copeland HE, Doherty KE, Naugle DE, Pocewicz A, Kiesecker JM (2009) Mapping Oil and Gas Development Potential in the US Intermountain West and Estimating Impacts to Species. PLoS ONE 4: e7400.

18. Kiesecker JM, Copeland H, Pocewicz A, Nibbelink N, McKenney B (2009) A Framework for Implementing Biodiversity Offsets: Selecting Sites and Determining Scale. BioScience 59: 77.84

19. Kiesecker JM, Evans JS, Fargione J, Doherty K, Foresman KR (2011) Win-Win for Wind and Wildlife: A Vision to Facilitate Sustainable Development. PLoS ONE 6: e17566.

20. Weber TC, Allen WL (2010) Beyond on-site mitigation: An integrated, multi-scale approach to environmental mitigation and stewardship for transportation projects. Landscape and Urban Planning 96: 240.256

21. Thorne J, Girvetz E, McCoy M (2009) Evaluating Aggregate Terrestrial Impacts of Road Construction Projects for Advanced Regional Mitigation. Environmental Management 43: 936.948

22. Kroeger T, Manalo P (2007) Economic Benefits Provided by Natural Lands: Case Study of California's Mojave Desert. Washington, D.C.: Defenders of Wildlife. 109 p.
23. Bunn D, Mummert A, Hoshovsky M, Gilardi K, Shank S (2007) California Wildlife: Conservation Challenges, California's Wildlife Action Plan. Sacramento, CA: California Department of Fish and Game. 597 p.
24. USFWS (2011) Ash Meadows National Wildlife Refuge website. U.S. Fish and Wildlife Service.
25. André J, Hughson D (2009) Sweeney Granite Mountains Desert Research Center: an Interview with Director Dr. Jim André. Mojave National Preserve: Science Newsletter.
26. Randall JM, Parker SS, Moore J, Cohen B, Crane L (2010) Mojave Desert Ecoregional Assessment. San Francisco, CA: The Nature Conservancy. 106 p.
27. Ball IR, Possingham HP, Watts M (2009) Marxan and relatives: Software for spatial conservation prioritisation. In: Moilanen A, Wilson KA, Possingham HP, editors. Oxford, U.K.: Oxford University Press.
28. BLM (2005) West Mojave Plan: A Habitat Conservation Plan and California Desert Conservation Area Plan Amendment. Moreno Valley, CA.
29. Clark County NV (2000) Final Clark County Multiple Species Habitat Conservation Plan and Environmental Impact Statement for Issuance of a Permit to Allow Incidental Take of 79 Species in Clark County, Nevada Las Vegas, Nevada.
30. Mehos MS, Owens B. Analysis of Siting Opportunities for Concentrating Solar Power Plants in the Southwestern United States; World Renewable Energy Congress VIII 2004; Denver, CO.
31. EIA US (2009) Energy Consumption by Sector and Source, United States, Early release reference. U.S. Department of Energy.
32. Pavlik B (2008) The California Deserts: An Ecological Rediscovery. Berkeley, CA: University of California Press. 344 p.
33. Lovich JE, Bainbridge D (1999) Anthropogenic Degradation of the Southern California Desert Ecosystem and Prospects for Natural Recovery and Restoration. Environmental Management 24: 309.326
34. Brown DE, Minnich RA (1986) Fire and Changes in Creosote Bush Scrub of the Western Sonoran Desert, California. American Midland Naturalist 116: 411.422
35. Wohlfahrt G, Fenstermaker LF, Arnone Iii JA (2008) Large annual net ecosystem CO_2 uptake of a Mojave Desert ecosystem. Global Change Biology 14: 1475.1487
36. Howard J, Merrifield M (2010) Mapping Groundwater Dependent Ecosystems in California. PLoS ONE 5: e11249.
37. USGS (2010) Landfire Existing Vegetation Type GIS dataset v.1.10. U.S. Geological Survey, U.S. Department of Interior.
38. Epps CW, Palsbøll PJ, Wehausen JD, Roderick GK, Ramey RR (2005) Highways block gene flow and cause a rapid decline in genetic diversity of desert bighorn sheep. Ecology Letters 8: 1029.1038
39. RETI (2010) Phase 2B Final Report. Sacramento, CA: RETI Coordination and Stakeholder Committees.

40. Jaccard P (1908) Nouvelles recherches sur la distribution florale. 44: 223–270. Bulletin de la Societe Vaudoise des Sciences Naturelles 44: 223.270
41. USFWS (1994) Desert tortoise (Mojave population) Recovery Plan. Poprtland, OR: U.S. Fish and Wildlife Service. 73 p.
42. Tracy CR, Averill-Murray R, Boarman WI, Delehanty D, Heaton J (2004) Desert Tortoise Recovery Plan Assessment. Reno: Desert Tortoise Recovery Plan Assessment Committee.
43. Perez R, Ineichen P, Moore K, Kmiecik M, Chain C (2002) A new operational model for satellite-derived irradiances: description and validation. Solar Energy 73: 307.317
44. NASA-JPL (2000) Shuttle Radar Topography Mission digital elevation model: 1 Arc Second. NASA.
45. ESRI (2009) ArcGIS v 9.3.1. Redlands, CA.
46. Lowry JH Jr, Ramsey RD, Boykin K, Bradford D, Comer P (2005) Southwest Regional Gap Analysis Project: Final Report on Land Cover Mapping Methods. USGS and RS/GIS Laboratory, Utah State University.
47. Calfire (2006) Multisource Land Cover dataset, fveg06. Calfire, Sacramento, CA.
48. RETI (2008) Phase 1B Final Report. Sacramento, CA: RETI Coordination and Stakeholder Committees.
49. Nussear KE, Esque TC, Inman RD, Gass L, Thomas KA (2009) Modeling Habitat of the Desert Tortoise (Gopherus agassizii) in the Mojave and Parts of the Sonoran Deserts of California, Nevada, Utah, and Arizona. 18 p.
50. BLM (2004) Mohave Ground Squirrel conservation areas: GIS dataset. Bureau of Land Management, California Desert District.
51. USGS (2010) Protected Areas Database (v 1.1). USGS.
52. Alvarado A, Rhyne I (2011) Proposed Method to Calculate the Amount of New Renewable Generation Required to Comply with Policy Goals: Draft Staff Paper. Sacramento, CA.
53. Miles SR, Goudey CB (1997) Ecological Subregions of California: Section and Subsection Descriptions. San Francisco, CA: USDA Forest Service Pacific Southwest Region.
54. SCBD (2008) Business 2010. A newsletter on business and biodiversity by the Secretariat of the Convention on Biological Diversity 3.

There are several supplemental files that are not available in this version of the article. To view this additional information, please use the citation on the first page of this chapter.

PART III

SOLAR TECHNOLOGY

CHAPTER 3

CARBON NANOTUBE SOLAR CELLS

COLIN KLINGER, YOGESHWARI PATEL, AND HENK W. CH. POSTMA

3.1 INTRODUCTION

Solar cells have great potential as an alternative energy source because of the enormous amount of available energy and its distributed nature that may enable a distributed power generation grid [1]. However, for solar energy to be cost-effective on a utility scale, the price of purchase, installation, operation and maintenance over the lifetime of a solar panel per kWh generated must compare favorably to current power generation technology, which for fossil-fuel based generation is 0.03–0.05$/kWh [2]. Improvements are being made to solar cells to 1) increase the efficiency, and 2) lower the price. For instance, solar concentrators are being developed that focus solar light reflecting off a large mirror on a solar cell with a smaller surface area. Multi-junction devices are being developed that use junctions between materials with different band gaps to capture a greater number of photons and limit loss of excess photon energy when the excited high-energy electron relaxes to the Fermi level.

Carbon Nanotube Solar Cells. © *Klinger C, Patel Y, and Postma HWC. PLoS ONE 7,5 (2012); doi:10.1371/journal.pone.0037806. Licensed under the Creative Commons Attribution 3.0 Unported License, http://creativecommons.org/licenses/by/3.0/.*

Gratzel cells [3], also known as Dye-Sensitized Solar Cells (DSSCs), offer a particularly interesting path to cost-effective solar power. By sacrificing some efficiency but offering a greater reduction in cost, the total price per kWh can be reduced considerably. While this initial argument for DSSCs is very compelling, it is worth noting that the current state-of-the art DSSCs have efficiencies that rival their solid-state counterparts [4]–[6]. Another advantage of DSSCs is that they operate well in low-light and overcast conditions. DSSCs typically consist of a transparent semiconducting film on conducting glass that functions as a photo-active electrode (figure 1a, top). A glass plate is coated with Pt and acts as the counter electrode (figure 1a, bottom). Light-sensitive dye molecules are adsorbed on a semiconducting material on another slide and the assembly is immersed in an electrolyte, typically iodide-triiodide (I^-/I_3^-). An incoming photon with energy hv excites an electron from the dye into the conduction band of the semiconductor and it migrates to the bottom electrode. The electrolyte reduces the dye, creating triiodide ($3I^- \rightarrow I_3^- + 2e^-$). The electrons follow the external circuit through the load to the counter electrode. The triiodide migrates through the electrolyte to the Pt electrode and gets reduced, thereby completing the circuit. The transparent semiconductor is typically made of nanoporous TiO_2. Using a nanoporous material significantly increases the surface area available for dye molecules but at the same time limits the electron migration rate. Different transparent semiconductors are being studied with higher mobility, such as nanowire-based electrodes [7], [8]. The liquid electrolyte is not very stable at the wide range of temperatures solar cells typically are exposed to, so high-mobility solids are being investigated as well [9], [10], culminating recently in a record 12% conversion efficiency [6]. Various dyes have been used in DSSCs, ranging from metal-free organic dyes [11] through highly efficient Ru-based organic dyes such as 'N3 dye' [12], [13] and 'black dye' [14]–[17] to engineered semiconductor quantum dots with a very high extinction coefficient [18]. C_{60} has been shown to work as a 'dye' as well [19], [20]. Carbon nanotubes (CNTs) [21], [22], offer a potentially cheaper and easier alternative to these materials. They are photo active, highly conductive, strong, and chemically inert. Carbon nanotubes can be synthesized in multiple ways such as chemical vapor deposition or laser ablation. The natural ratio of as-synthesized carbon nanotubes is 2/3 semiconducting to 1/3 metallic.

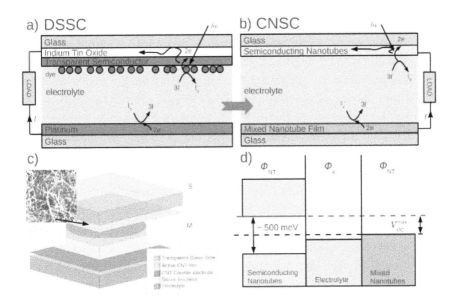

FIGURE 1: Carbon nanotube solar cells; comparison to Dye-Sensitized Solar Cells (DSSC), construction, and energeticts. a) DSSC. b) Carbon Nanotube Solar Cell, CNSC. c) Layout of a CNSC. The top and bottom glass slides are covered in carbon nanotube films which are electrically connected by the iodide-triiodide electrolyte that is contained by the silicone separator. The top film is the photoactive electrode, while the bottom electrode is the counter electrode. The inset is an Atomic Force Micrograph of the height of a 2×2 m section of a carbon nanotube film. d) Band diagram of the CNSC.

Here, we present proof-of-concept solar cells that are entirely made of carbon nanotubes, carbon-nanotube-based solar cells (CNSCs, figure 1b). They are a variation on the DSSC, and potentially offer many advantages beyond DSSCs. 1) No Dye. As these cells use semiconducting CNTs for photo conversion, they do not rely on dyes, which may bleach, severely limiting the useful life of DSSCs. 2) No Pt. Pt is often used as counter electrodes and their use in DSSCs represent an undesirable reliance on noble metals which may inhibit the use of DSSCs on a large, i.e. utility, scale. In addition, Pt has been reported to degrade due to the contact with the electrolyte [23]. Carbon nanotubes, in contrast, are chemically inert, and indeed show promising characteristics as counter electrodes [24]–[27]. 3)

No In. As the carbon nanotube film itself is a transparent conductor, the use of a conducting coating made of, e.g. InSnO, is not required, eliminating the need for the exceedingly rare Indium. 4) The application of carbon nanotubes to the glass slides is a low temperature spray-coating process. In addition, these CNSCs multiply the advantages offered by DSSCs over single and multi-junction solar cells that require high-grade semiconductors and clean-room manufacturing. The use of low-grade materials and resulting projected significant reduction in cost of manufacturing potentially offsets the limited efficiency of these cells when relating the energy produced per dollar spent in manufacturing and installation.

In addition to CNT-only cells, we report on effiency improvement strategies, using different assembly techniques and using graphite (graphenium) counter electrodes. Graphite has no band gap, is extremely pliable, robust, and provides the ability to shrink the distance between it and the active semiconducting electrode. The cost, relative abundance, ease of introduction into the cell, and lack of need for spray deposition render graphite an attractive counter electrode material.

3.2 RESULTS

We present experimental demonstration of power generation obtained under ambient conditions at solar noon (see Methods section for details) of two types of cells. 1) CNT-only cells: cells are built with identical geometry but different CNT film compositions and thickness. This highlights how film composition affects cell performance (figure 2a). 2) Optimized cells: cells are built with the same CNT film thickness and composition, but with differences in construction techniques to isolate its role in cell efficiency (figure 2b–d).

3.2.1 CNT-ONLY CELLS (FIGURE 1A)

The photocurrent decreases linearly with increasing cell potential applied to the load V (figure 3a). We extract the open-circuit voltage V_{oc} by extrapolating the I–V characteristic to I = 0 and the short-circuit current I_{sc}

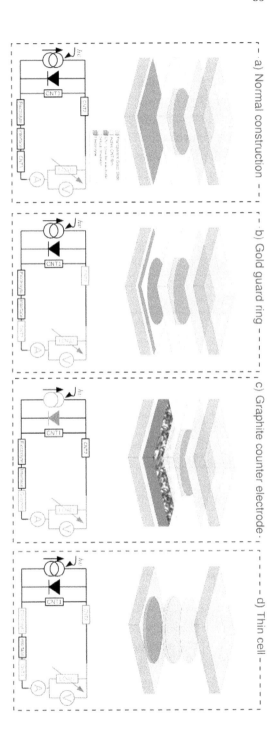

FIGURE 2: Layout of DSSCs and equivalent circuits. Both basic DSSCs (a) and tested optimization strategies are (b–d) are depicted. The circuit diagram is modeled after [41]. The alternative construction techniques lead to changes in the cell's electrical model, which are highlighted.

by extrapolating to V = 0 (table 1). Both I_{sc} and V_{oc} of the enriched mixture cells increase with decreasing CNT coverage of the semiconducting active electrode. Similarly, the high-density cell of the regular mixture of nanotubes has a lower I_{sc} and V_{oc} than the low-density cell. The power transfer curves (figure 3b) show a peak power transfer of P_{max} that occurs when the impedance of the load reaches R_{max}. The low-density enriched as well as the low-density regular cells deliver more power to the load than their high-density counterparts. This is consistent with both I_{sc} and V_{oc} being larger.

TABLE 1: Parameters of CNSCs.

		Cell Type	$R_\square$	V_{oc}	I_{sc}	P_{max}	R_{mac}	n_s	n_m
			$k\Omega/\square$	mV	nA	nW	$M\Omega$	a.u.	a.u.
■		Enriched High Density	3.4	43.6	47.0	0.57	0.71	521	58
◆	(black)	Enriched Low Density	62.7	208.5	243.4	11.50	0.82	121	13
◆	(blue)	Regular High Density	30.2	9.3	4.4	0.01	5.55	36	18
▲	(magenta)	Regular Low Density	50.3	21.4	19.0	0.17	2.45	29	15
▲	(green)	Enriched Medium Density	46.1	154.1	91.2	5.32	3.01	136	15

3.2.2 OPTIMIZED CELLS (FIGURE 2B–D)

We have studied cells with different construction techniques, using CNT electrodes from the same batch. Similar techniques were employed for data analysis as above (figure 4, table 2). The power transfer curves (figure 4b) show a peak power transfer P_{max} at $R = R_{max}$. Gold Guard Ring. The presence of the gold guard ring increases I_{sc} by a factor ~2.5, while V_{oc} remains approximately constant. R_{max} is lower by ~3.5 and P_{max} is ~2 times greater. Graphite Counter Electrode. Both V_{oc} and I_{sc} are greater than the normally constructed cell and P_{max} is ~12 times greater. Thin Cell. When

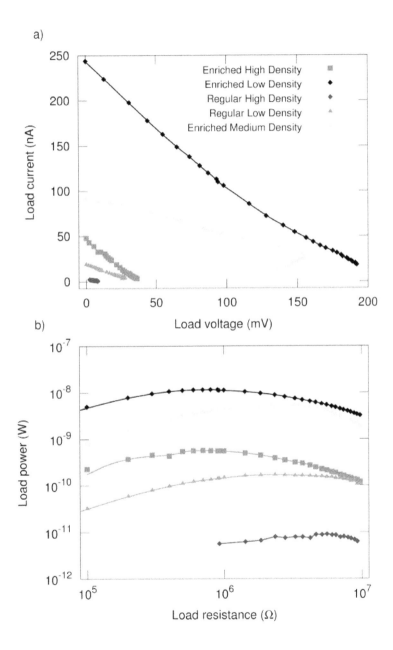

FIGURE 3: Electrical characteristics of the CNSCs. The extracted parameters are presented in table 2. a) I–V characteristics of the cells as indicated. b) Power delivered to the load for all cells as described in the legend for a).

the enriched side is facing the incident solar radiation ("up"), the power is slightly larger than when the regular side is facing the incident radiation ("down"). Both V_{OC} and I_{SC} are lower by factors of ~5 and ~3, which in itself is undesirable. However, optimum power transfer occurs at a much lower resistance.

TABLE 2: Optimized CNSCs.

	Cell Type	V_{OC}	I_{SC}	P_{max}	R_{max}
		mV	nA	nW	MΩ
■	Normal Construction	102.7	93.7	5.59	2.85
•	Gold Guard Ring	129.5	249.5	10.46	0.81
*	Graphite Counter Electrode	438.1	733.2	65.48	0.94
×	Thin Semiconducting Side Up	22.0	38.3	0.22	0.47
+	Thin Semiconducting Side Down	20.9	32.0	0.20	0.58

3.3 DISCUSSION

3.3.1 PHOTOCURRENT GENERATION AND CELL VOLTAGE

The linear I–V characteristic phenomenologically indicates the source is purely resistive, and maximum power occurs when the load and source impedance are equal. A figure of merit for solar cells that describes how close its I–V characteristic is to the ideal shape is the fill factor FF which is defined as the ratio of P_{max} to the maximum power available with the corresponding ideal cell, $FF \equiv P_{max} / (V_{OC} I_{SC})$. It ranges from 0 to 1, where 1 indicates an ideal cell. Ideal cells can supply a constant voltage independent on the load resistance up to the maximum current, when the voltage drops quickly to 0. Deviations from the ideal fill factor of 1 are usually due to parasitic resistances, such as shunt and series resistances. Shunt resistances affect behavior in the I–V characteristic close to I_{SC}, while series resistances affect performance close to V_{OC}. For our cell, FF ≈ 0.25. We argue that nanotube resistances 1–3 (figure 2) are responsible for this.

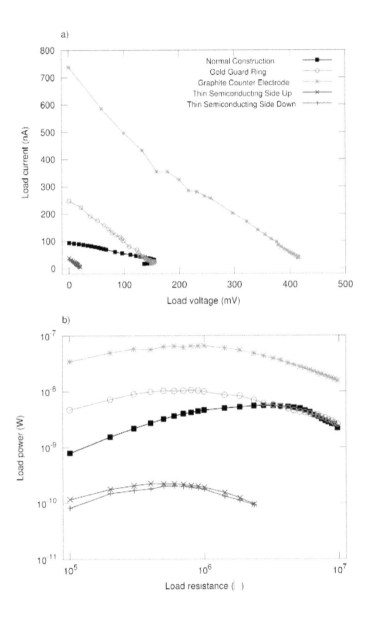

FIGURE 4: Optimization strategies for CNSCs. The extracted parameters are presented in table 1. a) I–V characteristics of the cells as indicated. b) Power delivered to the load.

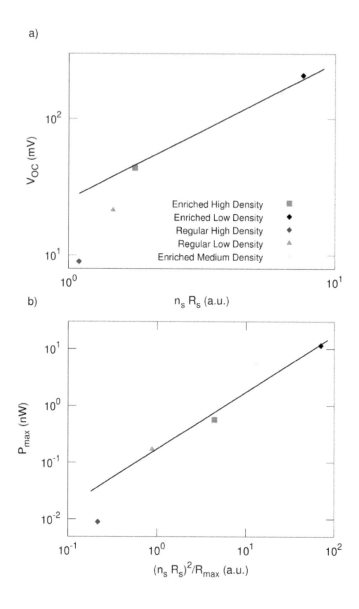

FIGURE 5: Scaling analysis of CNSC performance. The CNSCs characteristics are determined by the metallic and semiconducting carbon nanotube densities, with symbols corresponding to the cells as in the legend for figure 2. a) $V_{OC} \propto n_s R_\square$. b) $P_{max} \propto (n_s R_\square)^2 / R_{max}$.

Effectively, it means that the diode in the circuit diagram can be neglected. To estimate the number of nanotubes, we use the measured sheet resistance presented in table 1. Our CNT films are in the percolation limit [28], [29]. We can therefore use the scaling of sheet resistance with number of CNTs to extract the deposited volume of nanotube dispersion V, via

$$R_\square \propto (V-V_c)^{-1.5} \tag{1}$$

where V_C is the critical volume that determines the onset of conduction [28]. The volume can then be used to extract the surface density of metallic n_m and semiconducting nanotubes n_s. We assume that the metallic nanotubes dominate the sheet conductance, since their conductance G_m is much greater than semiconducting nanotubes G_s. This assumption holds provided the conductance ratio G_m/G_s of metallic to semiconducting nanotubes exceeds the semiconducting to metallic abundance ratio n_s/n_m. Single-molecule conductance studies of nanotubes indicate a conductance ratio of $G_m/G_s \approx 20$, [30], [31] which supports our assumption that metallic nanotubes dominate the sheet conductance. We anticipate that for more enriched semiconducting films than studied here, a more detailed analysis will be required that takes into account the nanotube-nanotube contact resistance as well [32], [33]. The current-generating capacity of our cells is proportional to the number of semiconducting nanotubes n_s. Combining both, the open-circuit condition corresponds to an ideal current source (I $\propto n_s$) connected to CNT1 ($R_{CNT1} \propto R_\square$) and the voltage developed across it will be

$$V_{OC} \propto n_s R_\square \tag{2}$$

and our data indeed approximately follows this scaling behavior (figure 5a). The outliers at low V_{OC} are CNT cells where both photoactive and counter electrode are coated with the same composition of carbon nanotubes. Both sides of the cell therefore create a photocurrent in opposite directions, but the light attenuation in the electrolyte breaks this symmetry and causes a directed current, albeit a smaller one and with a smaller voltage. The enriched cells further tilts the balance in favor of the photoactive side, leading to a V_{OC} that is closer to that expected from the amount of nanotube material deposited on the active side alone.

Our cells have rather large output impedances and cannot maintain constant voltage over a larger range of load impedance. The cell output resistance can be reduced considerably by changing the aspect ratio of the cell, or connecting many cells in parallel. The output voltage can be held constant by a voltage-regulation circuit. However, there are many applications that do not require a low output impedance and would therefore work well with CNSCs, e.g. driving an LCD display or an E-Ink screen.

3.3.2 MAXIMUM OPEN CIRCUIT VOLTAGE

The band gap of semiconducting carbon nanotubes is related to the nanotube's diameter d through

$$E_g = 2\gamma_0(a_{cc}/d) \tag{3}$$

where $\gamma_0 = 2.45eV$ is the nearest-neighbour overlap integral and a_{cc} is the carbon-carbon distance [30], [34]–[36]. Note that the band gap of semiconducting nanotubes does not depend on the chiral angle. The band gap for a 1.5 nm nanotube, the average diameter of our nanotube material, therefore amounts to ~500meV. The band diagram is drawn in figure 1d. The 'work function' of the electrolyte is $\Phi_e = 4.85eV$ [37], while the work function for carbon nanotubes is $\Phi_{NT} = 4.5eV$ [36]. We therefore expect the maximum attainable open circuit voltage V_{OC}^{max} ~200 – 290 mV, where the range indicates variations due to the diameter. We observe a maximum voltage of ~200mV. This is expected as nanotubes with a slightly smaller band gap will 'short out' the effect of nanotubes with a slightly larger bandgap.

3.3.3 SOLAR POWER GENERATION

The maximum power delivered to the load is a function of V_{OC} as well as the other resistances in the cell. The composite resistance of the cell is measured by determining at what value R_{max} of R_L maximum power transfer occurs. Since we can model our cells as a voltage source with source voltage $V_{OC} \propto n_s R_\square$, the maximum power available is expected to be $P_{max} \propto V_{OC}^2/R_{max}$, or

$$P_{max} \propto (n_s R_\square)^2/R_{max} \qquad (4)$$

and indeed the maximum power appears to follow this behavior approximately (figure 5b).

In summary, both P_{max} and V_{OC} behave according to our model that describes the role of film's resistive properties on cell performance.

3.3.4 OPTIMIZED CELL DESIGNS

The gold guard ring causes 1) an increase in I_{SC}, 2) an increase in P_{max}, 3) a decrease in R_{max}, and 4) hardly any change in V_{OC}. We believe this is due to a reduction of the resistance of the nanotube film in contact with the silicone insulator (CNT2 and CNT3, figure 2). The pressure of the insulator on the nanotube film as well as residual shear force during assembly may cause a perturbation of the nanotube film. In addition, the gold lowers the resistance of that part of the nanotube film. Both effects combined act to lower $R_{CNT2} + R_{CNT3}$. As the open-circuit voltage is independent of $R_{CNT2} + R_{CNT3}$, the open-circuit voltage should be unaffected by this improvement in design, and indeed we observe V_{OC} (Normal) $\sim V_{OC}$ (Gold Guard Ring). The reduction of R_{max} is also explained by the reduction of $R_{CNT2} + R_{CNT3}$, and that, in turn, explains the increase of both I_{SC} and P_{max}.

The employment of a graphite counter electrode instead of a carbon-nanotube counter electrode not only improves I_{SC}, but also V_{OC}. The increase in V_{OC} is due to the use of graphite instead of carbon nanotubes. The increase in I_{SC} is due to the lower sheet resistance of the graphite as compared to a carbon-nanotube film. The magnitude of the improvement is similar to that accomplished with the gold guard ring improvement, as the maximum power transfer occurs at approximately the same load resistance [R_{max} (Gold Guard Ring) $\sim R_{max}$ (Graphite Counter Electrode)]. Graphite is preferred over gold, naturally, to reduce cell cost. Both effects combine to increase the power output of the cell by a factor of ~12.

The reduction of the distance between active and counter electrode for the thin cells can reasonably be expected to lower the resistance of the electrolyte. In addition, as these cells were constructed without a silicone separator, we observe that the maximum power transfer occurs at a much lower resistance. This is due to the absence of the disruptive effect of the silicone separator, which role was elucidated by the study of the gold guard ring device above. The reduction of the electrolyte chamber thickness also has an adverse effect. The electrolyte absorbs less solar radiation than with thicker devices. Therefore, both the enriched (photo active) side, as well as the regular mixture side create a photo current. However, both sources have opposite polarities, causing the effective open-circuit voltage to be reduced as we indeed observe.

3.3.5 EFFICIENCY

The average solar flux during testing was 770 W/m², and the greatest solar power generation was attained with the graphite counter electrode and enriched medium-density CNT active electrode. The efficiency of that cell was 1.8×10^{-5}. Compared to the all-CNT construction, an improvement of more than a factor 10 was attained. If a cell were constructed with the graphite counter electrode and the low-concentration CNT enriched active electrode, an increase of power by a factor 2 is anticipated. This can be deduced by comparison of the medium density enriched cell to the low density enriched cells with the regular construction. As the graphite counter electrode lowered the output resistance by a factor ~3, the power output may be larger by a factor 3 as well. Further improvements may be obtained by changing the aspect ratio of the solar cell. In the design reported here, we used effectively square films.

Changing the cell design by making the cells wider, will lower the resistance further. An aspect ratio of 10 can then reduce the film resistance by a factor of 10, causing a reduction of R_{max}, which will improve P_{max}. Our thin cell results indicate that the largest resistance is due to the nanotube film, we therefore believe the efficiency increase with this improvement may be as large as 10-fold. We believe the greatest efficiency increase may be obtained by using CNT source material with a greater fraction of semiconducting nanotubes. The films we used had 90% semiconducting nanotubes and 10% metallic nanotubes. As we argued above, the semiconducting nanotubes provide the photo-generated current, but the metallic nanotubes short the load. If one were to use 99% semiconducting films, the amount of nanotubes could be increased by a factor 10, while still maintaining the same number of metallic nanotubes. As metallic nanotubes are more conductive than semiconducting nanotubes, we assume that the number of semiconducting nanotubes can be increased by this factor 10 without affecting $R_{\square}$. Future generation cells can then reasonably be expected to deliver 100 times more power, due to the increase of n_s by a factor 10 (equation 4). However, at a certain abundance factor of semiconducting to metallic nanotubes, this argument will not hold any longer. Combining all of these improvements may lead to an efficiency of 0.8–5%, where the lower bound is a conservative estimate that every improvement will only contribute half we argued above. We hope the studies reported here will motivate further development of methods to create highly-enriched semiconducting CNT source material cost effectively at a large scale.

3.4 MATERIALS AND METHODS

3.4.1 CELL CONSTRUCTION

The enriched CNSCs (figure 1) consist of a transparent glass slide covered with an enriched mixture of 90% semiconducting and 10% metallic nanotubes (IsoNanotubes-S 90% Powder, Nano Integris Inc.). These nanotubes have a diameter of 1.2–1.7 nm and a length of 0.1–4 m. Below this is a silicone insulator with a hole filled with electrolyte (iodide-triiodide, Solaronix). The electrolyte is in contact with both carbon nanotube films and acts to reduce the photo-active side as well as close the electrical circuit at the counter electrode. At the bottom is a glass slide covered with a regular mixture of 2/3 semiconducting and 1/3 metallic nanotubes that acts as a counter electrode (Unidym, lot PO-325, formerly Carbon Nanotechnologies Inc.). These nanotubes have a diameter of 0.8–1.2 nm and a length of 100–1000 nm. Dispersion of carbon nanotubes were made by ultrasonic agitation in 1,2-dichloroethane for 1 h for the regular mixture and 4 h for the enriched mixture. The dispersion was spray painted with an air brush onto glass substrates in a vented cylindrical enclosure. The slides were rotated while spraying to obtain uniform coverage. Subsequently, the glass slides were heated on a hot plate to evaporate any residual solvent. The resulting film is similar to the well-known bucky paper and it has metallic properties [38]–[40]. The final solar cell has an exposed surface area of ~4.8 mm^2 with a distance of ~2.5 mm between the electrodes. The glass slides are 1 mm thick and did not have a conducting coating prior to carbon nanotube application.

3.4.2 GOLD GUARD RING (FIGURE 2B)

A mask the size of the opening containing the electrolyte was placed onto the glass slide after CNT deposition, followed by Au deposition. This procedure prevents degration of the metal due to the electrolyte contact.

3.4.3 GRAPHITE CELL (FIGURE 2C)

A cell using graphite (graphenium) as the counter electrode was created. The graphite cell counter electrode construction consists of the same steps for deposition of semiconducting CNTs. A PDMS (Slygard 184 Silicone Elastomer, Dow Corning Corp.) plastic mold with a circular depression was created to house pieces of graphite of different heights. A wire is placed through the PDMS at the height of the bottom of the depression. After graphite deposition the PDMS was filled with electrolyte and the active semiconducting electrode was placed on top of the cell.

3.4.4 THIN CELL (FIGURE 2D)

A cell with a separation of about ~0.65 mm between the active 90% semi-conducting electrode and a regular CNT counter electrode was created. A piece of 1 mm thick glass was locally machined to create a central depression with a connection to a ramped section. The glass was cleaned and masked in the non-machines areas. The glass was sprayed with the 2/3 semiconducting and 1/3 metallic nanotubes mixture. A second piece of glass was masked with the same pattern as the machined glass piece and sprayed with 90% semiconducting CNTs. The two nanotube electrodes were connected to external electrodes and the cell was filled with electrolyte and sealed with liquid silicone sealant. The liquid silicone was allowed to dry and harden creating a seal. The thin cell has an exposed surface area of ~48.9 mm^2 with a distance of 0.55–0.75 mm between the electrodes.

3.4.5 CNT FILM PREPARATION AND CHARACTERIZATION

The spray-painted CNT slides were imaged with an Atomic Force Microscope (Dual-Scan AFM, Pacific Nanotechnology, USA) to determine the coverage (figure 1). In addition, a probe station was used to measure the sheet resistance $R_\square (\Omega/\square)$ of the CNT films, by analyzing the distance dependence of the two-terminal resistance as a function of probe separation L on a semilog scale and performing a least-squares fit to

$$R = (R_\square/\pi)\log(L/d) \tag{5}$$

where d is the probe tip diameter. The counter electrodes used in this study were all obtained from the same batch in order to ensure uniformity and their sheet resistance was $R_\square \approx 50k \ \Omega/\square$.

3.4.6 SOLAR POWER GENERATION MEASUREMENTS

The assembled cells were connected to a load resistor that was varied from 0 to 10 MΩ through a current amplifier and the voltage V across and current I through it are measured as a function of R_L (figure 2). The cells were pointed straight at the sun and were measured in Northridge, CA (visibility: 10 miles, Latitude = 34N) at solar noon from Dec 2010 through April 2011. The sun's altitude β was between 32.5 and 65.7, yielding an air mass of AM = $1/\sin\beta \approx 1.7$. The average solar flux was 770 W/m^2. To minimize the effect of variability in solar conditions and cell assembly, devices were fabricated with large variations in carbon nanotube concentrations to highlight its effect on cell performance.

CNTs were created in batch operations, providing the ability to test various parameters and the resistances of electrodes used in experiments.

REFERENCES

1. Lewis NS (2007) Toward Cost-Effective solar energy use. Science 315: 798–801.
2. Goldemberg J, Johansson TB, Anderson D (2004) World energy assessment: overview 2004 update. United Nations.
3. Gratzel M (2004) Conversion of sunlight to electric power by nanocrystalline dye-sensitized solar cells. Journal of Photochemistry and Photobiology A 164: 3–14.
4. Gao F, Wang Y, Zhang J, Shi D, Wang M, et al. (2008) A new heteroleptic ruthenium sensitizer enhances the absorptivity of mesoporous titania _lm for a high efficiency dye-sensitized solar cell. Chemical Communications 2635–2637.
5. Shi D, Pootrakulchote N, Li R, Guo J, Wang Y, et al. (2008) New efficiency records for stable Dye-Sensitized solar cells with Low-Volatility and ionic liquid electrolytes. The Journal of Physical Chemistry C 112: 17046–17050.
6. Yella A, Lee H, Tsao HN, Yi C, Chandiran AK, et al. (2011) Porphyrin-Sensitized solar cells with cobalt (II/III)Based redox electrolyte exceed 12 percent efficiency. Science 334: 629–634.
7. Baxter JB, Aydil ES (2005) Nanowire-based dye-sensitized solar cells. Applied Physics Letters 86: 053114.
8. Law M, Greene LE, Johnson JC, Saykally R, Yang P (2005) Nanowire dye-sensitized solar cells. Nature Materials 4: 455–459.
9. Bach U, Lupo D, Comte P, Moser JE, Weissortel F, et al. (1998) Solid-state dye-sensitized mesoporous TiO 2 solar cells with high photon-to-electron conversion efficiencies. Nature 395: 583–585.

10. Saito Y, Kitamura T, Wada Y, Yanagida S (2002) Poly (3, 4-ethylenedioxythiophene) as a hole conductor in solid state dye sensitized solar cells. Synthetic Metals 131: 185–187.

11. Horiuchi T, Miura H, Sumioka K, Uchida S (2004) High efficiency of Dye-Sensitized solar cells based on Metal-Free indoline dyes. Journal of the American Chemical Society 126: 12218–12219.

12. Nazeeruddin MK, Kay A, Rodicio I, Humphry-Baker R, Mueller E, et al. (1993) Conversion of light to electricity by cis-X2bis(2,2'-bipyridyl-4,4'-dicarboxylate) ruthenium(II)charge-transfer sensitizers (X = cl-, br-, i-, CN-, and SCN-) on nanocrystalline titanium dioxide electrodes. Journal of the American Chemical Society 115: 6382–6390.

13. Kohle O, Ruile S, Gratzel M (1996) Ruthenium(II) Charge-Transfer sensitizers containing 4,4-Dicarboxy-2,2-bipyridine. synthesis, properties, and bonding mode of coordinated thio- and selenocyanates. Inorganic Chemistry 35: 4779–4787.

14. Nazeeruddin MK, Pchy P, Graetzel M (1997) Efficient panchromatic sensitization of nanocrystalline TiO 2 films by a black dye based on a trithiocyanatoruthenium complex. Chemical Communications 1997: 1705–1706.

15. Aiga F, Tada T (2003) Molecular and electronic structures of black dye; an efficient sensitizing dye for nanocrystalline TiO2 solar cells. Journal of Molecular Structure 658: 25–32.

16. Boschloo G, Lindstrom H, Magnusson E, Holmberg A, Hagfeldt A (2002) Optimization of dye-sensitized solar cells prepared by compression method. Journal of Photochemistry & Photobiology, A: Chemistry 148: 11–15.

17. Wang ZS, Yamaguchi T, Sugihara H, Arakawa H (2005) Significant efficiency improvement of the black dye-sensitized solar cell through protonation of TiO2 films. Langmuir 21: 4272–4276.

18. Vogel R, Hoyer P, Weller H (1994) Quantum-Sized PbS, CdS, Ag2S, Sb2S3, and Bi2S3 particles as sensitizers for various nanoporous Wide-Bandgap semiconductors. The Journal of Physical Chemistry 98: 3183–3188.

19. Li C, Mitra S (2007) Processing of fullerene-single wall carbon nanotube complex for bulk hetero-junction photovoltaic cells. Applied Physics Letters 91: 253112.

20. Li C, Chen Y, Wang Y, Iqbal Z, Chhowalla M, et al. (2007) A fullerenesingle wall carbon nanotube complex for polymer bulk heterojunction photovoltaic cells. Journal of Materials Chemistry 17: 24062411.

21. Iijima S, Ichihashi T (1993) Single-shell carbon nanotubes of 1-nm diameter. Nature 363: 603–605.

22. Bethune DS, Klang CH, de Vries MS, Gorman G, Savoy R, et al. (1993) Cobalt-catalysed growth of carbon nanotubes with single-atomic-layer walls. Nature 363: 605–607.

23. Koo B, Lee D, Kim H, Lee W, Song J, et al. (2006) Seasoning effect of dye-sensitized solar cells with different counter electrodes. Journal of Electroceramics 17: 79–82.

24. Trancik JE, Barton SC, Hone J (2008) Transparent and catalytic carbon nanotube films. Nano Letters 8: 982–987.

25. Ramasamy E, Lee WJ, Lee DY, Song JS (2008) Spray coated multi-wall carbon nanotube counter electrode for tri-iodide reduction in dye-sensitized solar cells. Electrochemistry Communications 10: 1087–1089.

26. Hwang S, Moon J, Lee S, Kim D, Lee D, et al. (2007) Carbon nanotubes as counter electrode for dye-sensitised solar cells. Electronics Letters 43: 1455–1456.
27. Kang M, Han Y, Choi H, Jeon M (2010) Two-step heat treatment of carbon nanotube based paste as counter electrode of dye-sensitised solar cells. Electronics Letters 46: 1509–1510.
28. Hu L, Hecht DS, Gruner G (2004) Percolation in transparent and conducting carbon nanotube networks. Nano Letters 4: 2513–2517.
29. Zhou Y, Hu L, Gruner G (2006) A method of printing carbon nanotube thin films. Applied Physics Letters 88: 123109.
30. Tans S, Verschueren A, Dekker C (1998) Room-temperature transistor based on a single carbon nanotube. Nature 393: 49–52.
31. Postma HWC, Teepen T, Yao Z, Grifoni M, Dekker C (2001) Carbon nanotube Single-Electron transistors at room temperature. Science 293: 76–79.
32. Yao Z, Postma HWC, Balents L, Dekker C (1999) Carbon nanotube intramolecular junctions. Nature 402: 273–276.
33. Postma HWC, de Jonge M, Yao Z, Dekker C (2000) Electrical transport through carbon nanotube junctions created by mechanical manipulation. Physical Review B 62: R10653–R10656.
34. Martel R, Schmidt T, Shea HR, Hertel T, Avouris P (1998) Single- and multi-wall carbon nanotube field-effect transistors. Applied Physics Letters 73: 2447.
35. Odom TW, Huang J, Kim P, Lieber CM (1998) Atomic structure and electronic properties of single-walled carbon nanotubes. Nature 391: 62–64.
36. Wildoer JWG, Venema LC, Rinzler AG, Smalley RE, Dekker C (1998) Electronic structure of atomically resolved carbon nanotubes. Nature 391: 59–62.
37. Tang Y, Lee C, Xu J, Liu Z, Chen Z, et al. (2010) Incorporation of graphenes in nanostructured TiO2 films via molecular grafting for Dye-Sensitized solar cell application. ACS Nano 4: 3482–3488.
38. Mickelson ET, Huffman CB, Rinzler AG, Smalley RE, Hauge RH, et al. (1998) Fluorination of single-wall carbon nanotubes. Chemical physics letters 296: 188–194.
39. Rinzler AG, Liu J, Dai H, Nikolaev P, Huffman CB, et al. (1998) Large-scale purification of single- wall carbon nanotubes: process, product, and characterization. Appl Phys A 67: 29–37.
40. Vigolo B, Penicaud A, Coulon C, Sauder C, Pailler R, et al. (2000) Macroscopic fibers and ribbons of oriented carbon nanotubes. Science 290: 1331–1334.
41. Han L, Koide N, Chiba Y, Islam A, Mitate T (2006) Modeling of an equivalent circuit for dye-sensitized solar cells: improvement of efficiency of dye-sensitized solar cells by reducing internal resistance. Comptes Rendus Chimie 9: 645–651.

CHAPTER 4

ENABLING GREATER PENETRATION OF SOLAR POWER VIA THE USE OF CSP WITH THERMAL ENERGY STORAGE

PAUL DENHOLM AND MARK MEHOS

4.1 INTRODUCTION

Falling cost of solar photovoltaic (PV) generated electricity has led to a rapid increase in the deployment of PV and projections that PV could play a significant role in the future U.S. electric sector. The solar resource itself is virtually unlimited compared to any conceivable demand for energy (Morton 2006); however, the ultimate contribution from PV could be limited by several factors in the current grid. One is the limited coincidence between the solar resource and normal demand patterns (Denholm and Margolis 2007a). A second is the limited flexibility of conventional generators to reduce output and accommodate this variable generation resource. At high penetration of solar generation, increased grid flexibility will be needed to fully utilize the variable and uncertain output from PV generation and shift energy production to periods of high demand or reduced solar output (Denholm and Margolis 2007b).

This chapter was originally published by the U.S. Department of Energy. U.S. Department of Energy, National Renewable Energy Laboratory, Enabling Greater Penetration of Solar Power via the Use of CSP with Thermal Energy Storage, by Paul Denholm and Mark Mehos, NREL/TP-6A20-52978 (November 2011). http://www.nrel.gov/csp/pdfs/52978.pdf (accessed 30 June 2014).

Energy storage provides an option to increase grid flexibility and there are many storage options available or under development. In this work we consider a technology now beginning to be deployed at scale—thermal energy storage (TES) deployed with concentrating solar power (CSP). PV and CSP are both deployable in areas of high direct normal irradiance such as the U.S. Southwest. From a policy standpoint, a simplistic approach to choosing a generation technology might be based simply on picking the option with the lowest overall levelized cost of electricity (LCOE). However, deployment based simply on lowest LCOE ignores the relative benefits of each technology to the grid, how their value to the grid changes as a function of penetration, and how they may actually work together to increase overall usefulness of the solar resource.

Both PV and CSP use solar energy to generate electricity, although through different conversion processes. A key difference between CSP and PV technologies is the ability of CSP to utilize high-efficiency thermal energy storage (TES) which turns CSP into a partially dispatchable resource. The addition of TES produces additional value by shifting solar energy to periods of peak demand, providing firm capacity and ancillary services, and reducing integration challenges. Given the dispatchability of CSP enabled by thermal energy storage, it is possible that PV and CSP are at least partially complementary. The dispatchability of CSP with TES can enable higher overall penetration of solar energy in two ways. The first is providing solar-generated electricity during periods of cloudy weather or at night. However a potentially important, and less well analyzed benefit of CSP is its ability to provide grid flexibility, enabling greater penetration of PV (and other variable generation sources such as wind) than if deployed without CSP.

In this work we examine the degree to which CSP may be complementary to PV via its use of thermal energy storage. We first review the challenges of PV deployment at scale with a focus on the supply/demand coincidence and limits of grid flexibility. We then perform a series of grid simulations to indicate the general potential of CSP with TES to enable greater use of solar generation, including additional PV. Finally, we use these reduced form simulations to identify the data and modeling needed for more comprehensive analysis of the potential of CSP with TES to provide additional flexibility to the grid as a whole and benefit all variable generation sources.

4.2 CHALLENGES OF SOLAR DEPLOYMENT AT HIGH PENETRATION

The benefits and challenges of large scale PV penetration have been described in a number of analyses (Brinkman et al 2011). At low penetration, PV typically displaces the highest cost generation sources (Denholm et al. 2009) and may also provide high levels of reliable capacity to the system (Perez et al 2008). Figure 1 provides a simulated system dispatch for a single summer day in California with PV penetration levels from 0% to 10% (on an annual basis). This figure is from a previous analysis that used

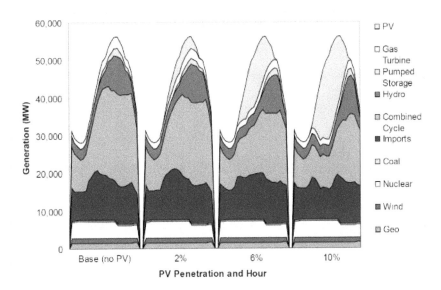

FIGURE 1: Simulated dispatch in California for a summer day with PV penetration from 0%–10% Note: Figure is modified from Denholm et al. (2008).

a production cost model simulating the western United States (Denholm et al. 2008). It illustrates how PV displaces the highest cost generation, and reduces the need for peaking capacity due to its coincidence with demand patterns.

At fairly low penetration (on an energy basis) the value of PV capacity drops. This can be observed in Figure 1 where the peak net load (normal load minus PV) stays the same between the 6% and 10% penetration curves. The net load in this figure is the curve at the top of the "Gas Turbine" area. Beyond this point PV no longer adds significant amounts of firm capacity to the system. Several additional challenges for the economic deployment of solar PV also occur as penetration increases. These are illustrated in Figure 2, which shows the results of the same simulation, except on a spring day. During this day, the lower demand results in PV displacing lower cost baseload energy. At 10% PV penetration in this simulation, PV completely eliminates net imports, and California actually exports energy to neighboring states.

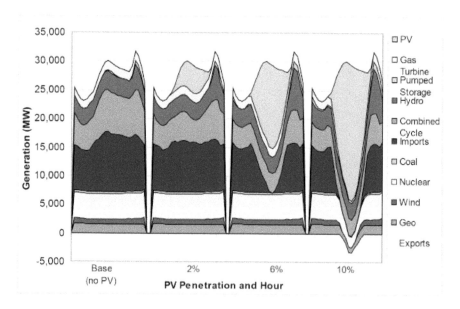

FIGURE 2: Simulated dispatch in California for a spring day with PV penetration from 0%–10% Note: Figure is modified from Denholm et al. (2008).

Several factors limit the ability of conventional generators to reduce output to accommodate renewable generation. These include the rate at which generators can change output, particularly in the evening when generators must increase output rapidly in a high PV scenario. This challenge is illustrated in Figure 3, a ramp duration curve for California covering an entire simulated year. This is the net load ramp rate (MW/hour) for all 8,760 hours in the simulated year ordered from high to low. In the no PV case, the maximum load ramp rate is about 5,000 MW/hour and a ramp rate of greater than 4,000 MW/hour occurs less than 100 hours in the simulated year. In the 2% PV case, the hourly ramps are actually smaller since PV effectively removes the peak demand (as seen in Figure 1). However at higher penetration, the ramp rates increase substantially, and in the 10% PV case the net load increases at more than 4,000 MW/hour more than 500 hours per year.

Another limitation is the overall ramp range, or generator turn-down ratio. This represents the ability of power plants to reduce output, which

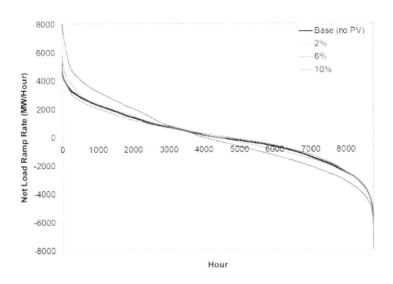

FIGURE 3: Ramp duration curve in California with PV penetration from 0% to 10%. Note: Figure is derived from Denholm et al. (2008).

is typically limited on large coal and nuclear units. Accommodating all of the solar generation as shown in Figure 2 requires nuclear generators to vary output which is not current practice in the U.S. nuclear industry. Most large thermal power plants cannot be turned off for short periods of time (a few hours or less), and brief shutdowns could be required to accommodate all energy generated during the period of peak solar output. The actual minimum load of individual generators is both a technical and economic issue—there are technical limits to how much power plants of all types can be turned down. Large coal plants are often restricted to operating in the range of 50%–100% of full capacity, but there is significant uncertainty about this limit (GE Energy 2010). Many plant operators have limited experience with cycling large coal plants, and extensive cycling could significantly increase maintenance requirements.

The ability to "de-commit" or turn off power plants may also be limited by the need to provide operating reserves from partially loaded power plants. As the amount of PV on the system increases, the need for operating reserves also increases due to the uncertainty of the solar resource, as well as its variability over multiple time scales.

Previous analysis has demonstrated the economic limits of PV penetration due to generator turn-down limits and supply/demand coincidence (Denholm and Margolis 2007a, Nikolakakis and Fthenakis 2011). Because of these factors, at high penetration of solar, increasing amounts of solar may need to be curtailed when its supply exceeds demand, after subtracting the amount of generation met by plants unable to economically reduce output due to ramp rate or range constraints or while providing operating reserves. Generator constraints would likely prevent the use of all PV generation in Figure 2. Nuclear plant operators would be unlikely to reduce output for this short period. Furthermore, PV generation may be offsetting other low or zero carbon sources. In Figure 2, PV sometimes displaces wind and geothermal generation, which provides no real benefit in terms of avoided fuel use or emissions.

While the penetration of solar energy is currently far too small to see significant impacts, curtailment of wind energy is an increasing concern in the United States (Wiser and Bolinger 2010). While a majority of wind curtailments in the United States are due to transmission limitations (Fink et al 2009), curtailments due to excess generation during times of low net load are a significant factor that will increase if grid flexibility is not enhanced. The resulting curtailed energy can substantially increase the levelized cost of energy (LCOE) from variable generators, because their capital costs must be recovered over fewer units of energy actually sold to the grid.

The ability of the aggregated set of generators to rapidly change output at a high rate and over a large range can be described as a grid's overall flexibility. Flexibility depends on many factors, including:

- Generator mix:Hydro and gas-fired generators are generally more flexible than coal or nuclear.
- Grid size: Larger grids are typically more flexible because they share a larger mix of generators and can share operating reserves and a potentially more spatially diverse set of renewable resources.
- Use of forecasting in unit commitment: Accurate forecasts of the wind and solar resources and load reduces the need for operating reserves.
- Market structure: Some grids allow more rapid exchange of energy and can more efficiently balance supply from variable generators and demand.
- Other sources of grid flexibility: Some locations have access to demand response, which can provide an alternative to partially-loaded thermal generators for provision of operating reserves. Other locations may have storage assets such as pumped hydro.

A comprehensive analysis of each flexibility option is needed to evaluate the cost-optimal approach of enhancing the use of variable generation. In this analysis, we consider the use of thermal energy storage. Previous analysis has demonstrated the ability of a wind and solar-based system to meet a large fraction of system demand when using electricity storage (Denholm and Hand 2011). A number of storage technologies are currently available or under development, but face a number of barriers to deployment including high capital costs efficiency related losses, and certain market and regulatory challenges. A number of initiatives are focused on reducing these barriers.

An alternative to storing solar generated electricity is storing solar thermal energy via CSP/TES. Because TES can only store energy from thermal generators such as CSP, it cannot be directly compared to other electricity storage options, which can charge from any source. However, TES provides some potential advantages for bulk energy storage. First, TES offers a significant efficiency advantage, with an estimated round trip efficiency in excess of 95% (Medrano et al. 2010). TES has the potential for low cost, with one estimate for the cost associated with TES added to a CSP power tower design at about $72/kWh-e (after considering the thermal efficiency of the power block).

4.3 SYSTEM MODEL

The purpose of this analysis is to explore the potential of CSP to provide grid flexibility and enable increased solar penetration in the Southwestern United States. To perform this preliminary assessment, we use the RE-Flex model, which is a reduced form dispatch model designed to examine the general relationship between grid flexibility, variable solar and wind generation, and curtailment (Denholm and Hand 2011). REFlex compares hourly load and renewable resources and calculates the amount of curtailment based on the system's flexibility, defined as the ability for generators to decrease output and accommodate variable generator sources such as solar and wind.

California is a likely candidate for large-scale deployment of both PV and CSP, and has strong solar incentive programs and a renewable portfolio standard. However, modeling California in isolation ignores the fact that California has strong transmission ties to neighboring states, including Arizona and southern Nevada, which have significant potential for solar energy. Currently, power exchanges between neighboring areas in the western United States are accomplished through bilateral contracts, and typically do not occur in real time. This analysis assumes the eventual availability of real-time power and energy exchanges across California, Arizona, New Mexico and Southern Nevada to allow sharing of solar resources. It also assumes that transmission is accessible to all generation sources on a short-term, non-firm basis. This "limiting case" allows for examination of the best technical case for solar deployment without market barriers or transmission constraints.

We began our simulations by evaluating the limits of PV, given flexibility limits of the existing grid. The simulations use solar, wind and load data for the years 2005 and 2006. Load data was derived from FERC Form 714 filings. For hourly PV production, we used the System Advisor Model

(SAM), which converts solar insolation and temperature data into hourly PV output (Gilman et al. 2008). Weather data for 2005 and 2006, was obtained from the updated National Solar Radiation Database (NSRDB) (Wilcox and Marion 2008). We assume that PV will be distributed in a mix of rooftop and central systems (both fixed and 1-axis tracking). Additional description of this mix, including geographical distribution is provided in Brinkman et al. (2011).

Because California has significant wind capacity installed and plans for more, we also consider the interaction between solar and wind generation. Simulated wind data for 2005 and 2006 for California/Southwest sites was derived from the datasets generated for the Western Wind and Solar Integration Study (WWSIS) (GE Energy 2010). We started with a base assumption that wind provides 10% of the region's energy based on the "In-Area –10% Wind" scenario from the WWSIS. These data sets were processed through the REFlex model to establish base relationships between grid penetration of PV, curtailment, and grid flexibility. The overall system flexibility was evaluated parametrically, starting with a base assumption that the system is able to accommodate PV over a cycling range of 80% of the annual demand range. This corresponds to a "flexibility factor" of 80%, meaning the aggregated generator fleet can reduce output to 20% of the annual peak demand (Denholm and Hand 2011). This value is based on the WWSIS study and corresponds roughly to the point where all on-line thermal units have reduced output to their minimum generation levels and nuclear units would require cycling. The actual flexibility of the U.S. power system is not well defined, and this value is not intended to be definitive, but is used to represent the challenges of solar and wind integration and the possible flexibility benefits of CSP/TES.

Figure 4 illustrates the framework for this analysis, showing the simulated dispatch over a 4-day period (April 7-10). It demonstrates a case where 10% of the annual demand is met by wind and 20% is met by solar. The figure shows both the simulated solar profile and its contribution to meeting load. Because of relatively low load during this period, PV generation exceeds what can be accommodated using the assumed grid flexibility limits. This typically occurs in the late morning, before the demand increases to its maximum in the afternoon. In these four days about 16% of all PV generation is curtailed and about 5% of the annual PV generation is curtailed.

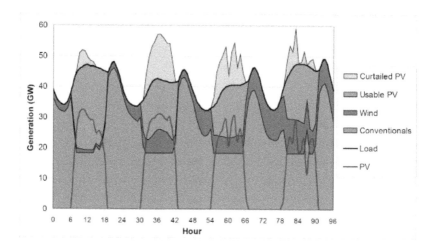

FIGURE 4: Simulated system dispatch on April 7-10 with 20% contribution from PV generation and resulting curtailment due to grid flexibility constraints

Figure 5 illustrates the average and marginal PV curtailment rates as a function of PV energy penetration for this initial scenario. It should be noted that the x-axis shows penetration of only solar PV. Because wind provides 10%, the total penetration of variable generation is 10% plus the penetration of solar. The average curve shows the total curtailment of all PV at a certain generation level. At the overall assumed system flexibility level, by the time PV is providing 22% of total demand, about 6% of all potential PV generation is curtailed.

The actual allocation of curtailment strongly influences the economics of PV and other variable generation. Figure 4 also shows the marginal curtailment rate, or the curtailment rate of the incremental unit of PV installed to meet a given level of PV penetration. If curtailment were assigned on an incremental basis at the point where PV is providing 22% of total demand, only about 50% of this additional PV would be usable, with the rest curtailed.

In this analysis we "assign" all incremental curtailment to solar, partially based on the federal production tax credit which incentivizes wind

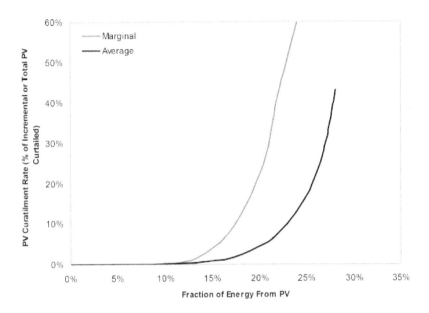

FIGURE 5: Marginal curtailment rates of PV in a base scenario in the southwestern United States assuming an 80% system flexibility

generation, while the primary federal incentive for solar is an investment tax credit that incentivizes installations but not generation. [15]

Curtailment of solar may also occur if wind is installed "first" and a "last in, first curtailed" rule applies. The actual allocation of curtailment is, and is likely to continue to be, a contentious issue. Regardless of allocations rules, increased grid flexibility will be needed to minimize curtailment if solar is expected to play a "primary" role in reducing fossil-fuel use in the electric sector.

The estimation of the marginal curtailment rate is important because it helps establish the optimal mix of generators serving various portions of the load. This can be observed in Figure 6, which translates curtailment into a cost of energy multiplier. This multiplier—equal to 1/(1-curtailment rate)—can be applied to the "base" LCOE of electricity generation (no curtailment). This represents how much more would need to be charged for electricity based on the impact of curtailment and the corresponding reduction in electricity actually provided to the grid.

Both the average and marginal multipliers are shown in Figure 6. The average multiplier is applied to all PV generators. The marginal multiplier is applied to the incremental generator, and is more important when determining the role of storage or other load-shifting technologies. For example, at the point where PV is providing 25% of the system's energy, the curtailment of all PV (average curtailment) is about 17% and the resulting cost multiplier is 1.2. If the base cost of PV is $0.06/kWh, the overall, system-wide cost of PV would be $0.06 x 1.2 or $0.072/kWh. This overall cost may be acceptable, but the costs are greater at the margin. For example, the last unit of PV installed to reach the 25% threshold has a curtailment rate of about 68% and a cost multiplier of 3.1. At a $0.06/kWh base price, this incremental unit of PV generation would have an effective cost

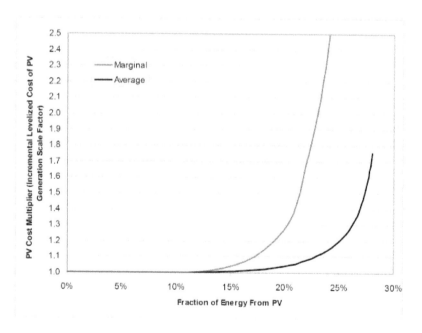

FIGURE 6: Impact of curtailment on PV LCOE multiplier in a base scenario in the southwestern United States assuming an 80% system flexibility

of more than $0.18 per kWh. This would likely result in examining options to both increase grid flexibility (to accommodate more PV with lower curtailment rates) and improve the solar supply/demand coincidence.

4.4 INCREASING SOLAR DEPLOYMENT USING CSP

While there are many options to increase grid flexibility, in this work we focus on the potential use of CSP with TES. Thermal storage extends the contribution of solar electricity generation by shifting generation to improve its coincidence with normal demand, and by improving system flexibility. The latter is accomplished by reducing constraints of ramping and minimum generation levels.

CSP was added to REFlex using hourly generation values produced by SAM. SAM uses the direct normal irradiance (DNI) to calculate the hourly electrical output of a wet-cooled trough plant (Wagner and Gilman 2011). The choice of technology was based primarily on data availability at the time of analysis as opposed to any presumption regarding CSP technology or economics. The results should be applicable to any CSP technology able to deploy multiple hours of thermal energy storage. For our base case, we assume 8 hours of storage and that the electrical energy produced by the plant can be dispatched with an effective 95% efficiency. In this initial analysis we did not consider the effects of part loading or multiple starts on plant efficiency. Distribution of locations was based on the study described by Brinkman et al. (2011).

Figure 7 illustrates the importance of dispatchability at high solar penetration. This scenario is identical to Figure 4, except PV provides 15% of annual demand and CSP meets 10% (so the contribution of solar technologies in total is greater in the PV/CSP case in Figure 7). The figure shows two CSP profiles. This first "non-dispatched CSP" is the output of CSP if it did not have thermal storage. It aligns with PV production, and would result in significant solar curtailment. The other curve is the actual dispatched CSP, showing its response to the net demand pattern after wind and PV generation is considered. It shows how a large fraction of the CSP energy is shifted toward the end of the day. In the first day, this ability to shift energy eliminates curtailment. On the other days, the wind and PV resources exceed the "usable" demand for energy in the early part of the day, resulting in curtailed energy even while the CSP plant is storing 100% of thermal energy. However, overall curtailment is greatly reduced. Solar technologies provide an additional 5% of the system's annual energy compared to the case in Figure 4, but the actual annual curtailment has been reduced to less than 2%, including the losses in thermal storage.

Figure 8 shows how the addition of CSP/TES can increase the overall penetration of solar by moving energy from periods of low net demand in the middle of the day to morning or evening. In this figure there is an equal mix of CSP and PV on an energy basis and the PV-only curves are identical to those in Figure 5.

Figure 8 demonstrates the importance of dispatchability to reduce curtailment and increase the overall penetration of solar via the ability to shift solar energy over time. However, the analysis to this point assumes that CSP and PV are complementary only in their ability to serve different parts of the demand pattern. We have not yet considered the additional benefits of CSP to provide system flexibility by replacing baseload generators and generators online to provide operating reserves.

The importance of system flexibility can be observed in Figure 4, where conventional generators must ramp up rapidly to address the decreased output of PV during peak demand periods. In order to meet this ramp rate and range (along with sufficient operating reserves) a significant number of thermal generators will likely need to be operating a part-load, creating a minimum generation constraint during periods of solar high output. This is represented by the flat line occurring in the middle of each day when the aggregated generator fleet is at their minimum generation point. Comparing the CSP/PV case in Figure 7 to the PV only case in Figure 4, we see that the CSP is dispatched to meet the peak demand in the late afternoon/ early evening, and the overall ramp rate and range is substantially reduced. In Figure 4 conventional generators need to ramp from about 18 GW to over 45 GW in just a few hours, while in Figure 7 the generators need to ramp from 18 GW to less than 30 GW.

Adding a highly flexible generator such as CSP/TES can potentially reduce the minimum generation constraint in the system. In the near term, this means that fewer conventional generators will be needed to operate at part load during periods of high solar output. In the longer term, the ability

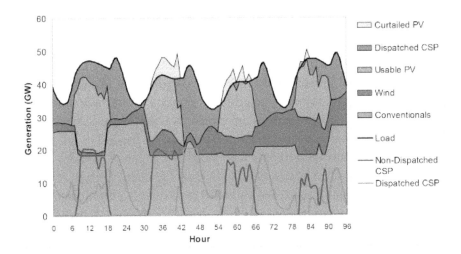

FIGURE 7: Simulated system dispatch on April 7-10 with 15% contribution from PV and 10% from dispatchable CSP

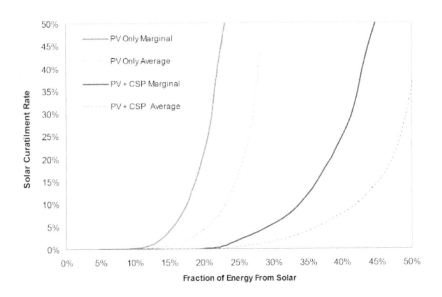

FIGURE 8: Curtailment of solar assuming an equal mix (on an energy basis) of PV and CSP

of CSP to provide firm system capacity could replace retiring inflexible baseload generators.

CSP plants with TES add system flexibility because of their large ramp rate and range relative to large baseload generators. Many CSP plants, both existing and proposed, are essentially small steam (Rankine-cycle) plants whose "fuel" is concentrated solar energy. Few of these plants are deployed, so it is not possible to determine their performance with absolute certainty. However, historical performance of the SEGS VI power plant provides some indication of CSP flexibility. Figure 9 provides a heat rate curve based on an hourly simulation model to assess the performance of parabolic trough systems, and validated by comparing the modeled output results with actual plant operating data (Price 2003). It indicates a typical operating range over 75% of capacity, with only a 5% increase in heat rate at 50% load. Figure 9 also provides historical data from small gas-fired steam plants which also indicates high ramp rate and range and fairly small decrease in efficiency at part load (about a 6% increase in heat

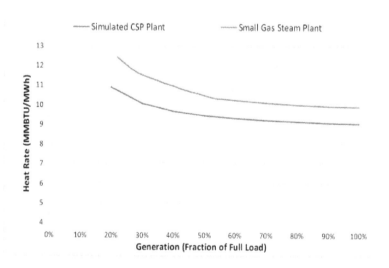

FIGURE 9: Part load heat rate of a CSP parabolic trough Rankine cycle power block and historic performance of small gas steam plants

rate at 50% load). These plants also often operate as low as 25% of capacity, although with lower efficiency. This provides a strong indication that CSP plants should be able to provide high flexibility.

The change in minimum generation constraints is dependent on both the flexibility of CSP plants and the flexibility of generators supplemented or replaced with CSP. As discussed previously, nuclear plants are rarely cycled in the United States, while coal plants are typically operated in the range of 50%-100%. Because it is not possible to determine the exact mix of generators that would be replaced in high renewables scenarios, we consider a range of possible changes in the minimum generation constraints resulting from CSP deployment. For example, deployment of a CSP plant which can operate over 75% of its capacity range could allow the de-commitment of a coal plant which normal operates over 50% of its range. In this scenario each unit of CSP could reduce the minimum generation constraint by 25% of the plant's capacity. This very simplistic assumption illustrates how the dispatchability of a CSP plant should allow for a lower minimum generation constraint.

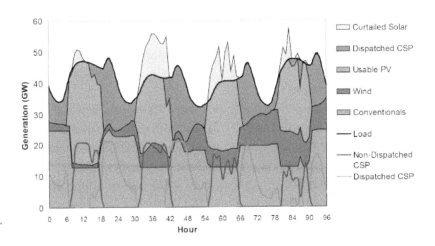

FIGURE 10: Simulated system dispatch on April 10-13 with 25% contribution from PV and 10% from dispatchable CSP where CSP reduces the minimum generation constraint

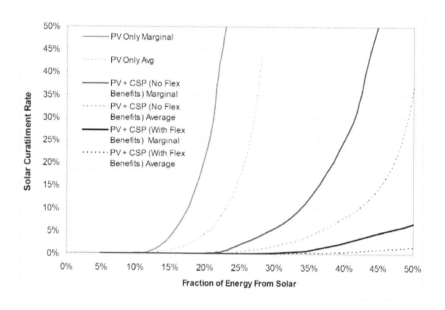

FIGURE 11: Curtailment of solar assuming an equal mix (on an energy basis) of PV and CSP and impact of CSP grid flexibility

Reducing this constraint should allow for greater use of wind and PV. As a result, as CSP is added, the system can actually accommodate more PV than in a system without CSP.

This is illustrated conceptually in Figure 10, which shows the same 4-day period as in Figures 4 and 7. CSP still provides 10% of the system's annual energy, but now we assume that the use of CSP allows for a decreased minimum generation point, and the decrease is equal to 25% of the installed CSP capacity. In this case about 21 GW of CSP reduces the minimum generation point from about 18 GW to 13 GW. This generation "headroom" allows for greater use of PV, and enough PV has been added to meet 25% of demand (up from 15% in Figure 7). As a result, the total solar contribution is now 35% of demand, significantly greater than the PV-only case shown in Figure 4, and total curtailment is less than the 6% rate seen in Figure 4. By shifting energy over time and increasing grid

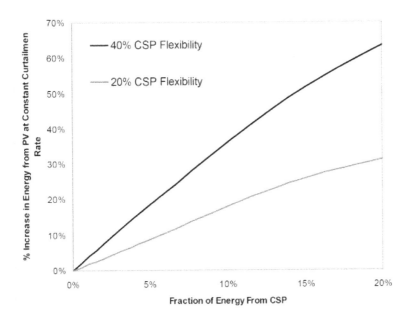

FIGURE 12: Increase in PV penetration as a function of CSP penetration assuming a maximum PV marginal curtailment rate of 20%. CSP flexibility is defined as the fraction of the CSP rated capacity that is assumed to reduce the system minimum generation constraint.

flexibility, CSP enables greater overall solar penetration AND greater penetration of PV.

Figures 11 and 12 show the potential overall impact of the flexibility introduced by CSP and the corresponding opportunities for increased use of PV. Figure 11 builds on Figure 8 by adding the flexibility benefits of CSP. The figure assumes that each unit of CSP reduces the minimum generation constraint by 25% of its capacity, and an equal mix of PV and CSP on an energy basis. In this case, the addition of CSP allows PV to provide 25% of the system's energy with very low levels of curtailment.

Figure 12 more directly illustrates the relationship between the reduction in minimum generation constraint and potential increase in PV penetration. The figure shows how much more PV could be incorporated at a constant marginal curtailment rate of 20% when CSP is added. In this scenario, the x-axis represents the fraction of annual system energy provided by CSP. Increased penetration of CSP results in a linear decrease in minimum generation constraints. The figure illustrates two CSP flexibility cases. In one, each unit of CSP reduces the minimum generation constraint by 20% of its capacity; in the other, the rate of reduction is 40%. These amounts are not meant to be definitive, but represent a possible impact of CSP in reducing minimum generation constraints.

Overall, this analysis suggests that CSP can significantly increase grid flexibility by providing firm system capacity with a high ramp rate and range and acceptable part-load operation. Greater grid flexibility could increase the contribution of renewable resources like solar and wind. This demonstrates that CSP can actually be complementary to PV, not only by adding solar generation during periods of low sun, but by actually enabling more PV generation during the day. This analysis also suggests a pathway to more definitively assess the ability of CSP to act as an "enabling" technology for wind and solar generation.

4.5 FURTHER QUANTIFYING THE BENEFITS OF CSP DEPLOYMENT

This analysis is a preliminary assessment of the potential benefits of CSP in providing grid flexibility using reduced form simulations with limited geographical scope and many simplifying assumptions. Gaining a more thorough understanding of how CSP can enable greater PV and wind penetration will require detailed production simulations using security-constrained unit commitment and economic dispatch models currently used by utilities and system operators. These simulations should consider the operation of the entire power plant fleet including individual generator characteristics and constraints, and the operation of the transmission system. The geographical footprint should cover the entire Western interconnect including possible transmission expansion to take advantage of greater spatial diversity of the wind and solar resources as well conventional generators.

To date, production simulations have not considered CSP operations in detail. Both the WWSIS and the first phase of the California 33% Renewable Portfolio Standard integration studies (CAISO 2011) included CSP, but assumed fixed schedules for CSP dispatch. This assumption limits CSP's ability to shift generation to when needed most and to provide grid flexibility to enable PV and wind. Future and ongoing studies, including the second phase of both the California study and the WWSIS will evaluate the benefits of TES in more detail. To perform these simulations, production cost models will need to include the ability of CSP to optimally dispatch the solar energy resource, and not rely on heuristics or schedules often used to estimate the operation of conventional storage plants such as pumped hydro. However, the ability to optimize CSP, including scheduling both its energy and ability to provide operating reserves, is limited by lack of certain data sets needed for a more detailed simulation. A greater understanding of the predictability and variability of the solar resource, including the sub-hourly variation and the effects of spatial diversity in mitigating variability, is needed. This data will also be needed to determine any required increase in operating reserves over various time scales as a function of solar penetration. In addition, more data is needed on the actual characteristics of CSP plants—those now being deployed and under development—including ramp rates, turn-down ratio, part-load efficiency, and start times under various conditions.

4.6 CONCLUSIONS

While it will be some time until solar technologies achieve very high penetrations in the U.S. grid, international experience in wind deployment demonstrates the importance of increasing overall grid flexibility. Key factors in improving grid flexibility include increasing the ramp range and rate of all generation sources and the ability to better match the supply of renewable resources with demand via increased spatial diversity, shiftable load, or energy storage. The use of thermal energy storage in concentrating solar power plants provides one option for increased grid flexibility in two primary ways. First, TES allows shifting of the solar resource to periods of reduced solar output with relatively high efficiency. Second is the inherent flexibility of CSP/TES plants, which offer higher ramp rates and ranges than large thermal plants currently used to meet a large fraction of electric demand. Given the high capacity value of CSP/TES, this technology could potentially replace a fraction of the conventional generator fleet and provide a more flexible generation mix. This could result in greater use of non-dispatchable solar PV and wind meaning CSP and PV may actually be complementary technologies, especially at higher penetrations.

The preliminary analysis performed in this work requires advanced grid simulations to verify the actual ability of CSP to act as an enabling technology for other variable generation sources. Complete production simulations using utility-grade software, considering the realistic performance of the generation fleet, transmission constraints, and actual CSP operation will be an important next step in evaluating the benefits of multiple solar generation technologies.

REFERENCES

1. Brinkman, G.L.; Denholm, P.; Drury, E.; Margolis, R.; Mowers, M. (2011). "Toward a Solar-Powered Grid—Operational Impacts of Solar Electricity Generation." IEEE Power and Energy (9); pp. 24–32.
2. California Independent System Operator (CAISO) (2011). "Track I Direct Testimony of Mark Rothleder on Behalf of the California Independent System Operator Corporation." Testimony for the Public Utilities Commission of the State of California, Order Instituting Rulemaking to Integrate and Refine Procurement Policies and Consider Long-Term Procurement Plans, Rulemaking 10-05-006, Submitted July 11.
3. Denholm, P.; Hand, M. (2011). "Grid Flexibility and Storage Required to Achieve Very High Penetration of Variable Renewable Electricity." Energy Policy (39); pp. 1817–1830.
4. Denholm, P.; Ela, E.; Kirby, B.; Milligan, M. (2010). The Role of Energy Storage with Renewable Electricity Generation. NREL/TP-6A2-47187. Golden, CO: National Renewable Energy Laboratory.
5. Denholm, P.; Margolis, R.M.; Milford, J. (2009). "Quantifying Avoided Fuel Use and Emissions from Photovoltaic Generation in the Western United States." Environmental Science and Technology (43); pp. 226–232.
6. Denholm, P.; Margolis, R.M.; Milford, J. (2008). Production Cost Modeling for High Levels of Photovoltaics Penetration. NREL/TP-581-42305. Golden, CO: National Renewable Energy Laboratory.

7. Denholm, P.; Margolis, R.M. (2008). "Land Use Requirements and the Per-Capita Solar Footprint for Photovoltaic Generation in the United States." Energy Policy (36); pp. 3531–3543.

8. Denholm, P.; Margolis, R.M. (2007a). "Evaluating the Limits of Solar Photovoltaics (PV) in Traditional Electric Power Systems." Energy Policy (35); pp. 2852–2861.

9. Denholm, P.; Margolis, R.M. (2007b). "Evaluating the Limits of Solar Photovoltaics (PV) in Electric Power Systems Utilizing Energy Storage and Other Enabling Technologies." Energy Policy (35); pp. 4424–4433.

10. Electric Power Research Institute (EPRI). (December 2010). "Electricity Energy Storage Technology Options: A White Paper Primer on Applications, Costs, and Benefits." 1020676. Palo Alto, CA: EPRI.

11. Fink, S.; Mudd, C.; Porter, K.; Morgenstern, B. (2009). Wind Energy Curtailment Case Studies: May 2008 - May 2009. SR-550-46716. Golden, CO: National Renewable Energy Laboratory.

12. GE Energy. (2010). Western Wind and Solar Integration Study. SR-550-47434. Golden, CO: National Renewable Energy Laboratory.

13. Gilman, P.; Blair, N.; Mehos, M.; Christensen, C.; Janzou, S.; Cameron, C. (2008). Solar Advisor Model User Guide for Version 2.0. TP-670-43704. Golden, CO: National Renewable Energy Laboratory.

14. Johnson, M. (2 March 2011). "Overview of Gridscale Rampable Intermittent Dispatchable Storage (GRIDS) Program." Washington, DC: U.S. Department of Energy.

15. Kearney, D.; Miller, C. (15 January 1998). "Solar Electric Generating System VI - Technical Evaluation of Project Feasibility." Los Angeles, CA: LUZ Partnership Management, Inc.

16. Kolb, G.; Ho, C.; Mancini, T.; Gary, J. (2011). Power Tower Technology Roadmap and Cost Reduction Plan. SAND2011-2419. Albuquerque, NM: Sandia National Laboratories.

17. King, J.; Kirby, B.; Milligan, M.; Beuning, S. (2011) Flexibility Reserve Reductions from an Energy Imbalance Market with High Levels of Wind Energy in the Western Interconnection. NREL/TP-5500-5233. Golden, CO: National Renewable Energy Laboratory.

18. Lefton, S.A.; Besuner, P. (2006). "The Cost of Cycling Coal Fired Power Plants." Coal Power Magazine, Winter 2006.

19. Medrano, M.; Gil, A.; Martorell, I.; Potau, X.; Cabeza, F. (2010). "State of the Art on High-Temperature Thermal Energy Storage for Power Generation. Part 2 – Case Studies." Renewable and Sustainable Energy Reviews (14); pp. 56–72.

20. Morton, O. (2006). "Solar Energy: A New Day Dawning?: Silicon Valley Sunrise." Nature (443); pp. 19–22.

21. Nikolakakis, T.; Fthenakis, V. "The Optimum Mix of Electricity from Wind- and Solar-Sources in Conventional Power Systems: Evaluating the Case for New York State." Energy Policy, (39); 6972-6980.

22. Perez, R.; Taylor, M.; Hoff, T.; Ross, J.P. (2008). "Reaching Consensus in the Definition of Photovoltaic Capacity Credit in the USA: A Practical Application of Satellite-Derived Solar Resource Data." IEEE Journal of Selected Topics In Applied Earth Observations And Remote Sensing (1:1); pp. 28–33.

23. Price, H. (2003). A Parabolic Trough Solar Power Plant Simulation Model. CP-550-33209. Golden, CO: National Renewable Energy Laboratory.
24. Sioshansi, R.; Denholm, P. (2010). "The Value of Concentrating Solar Power and Thermal Energy Storage." IEEE Transactions on Sustainable Energy (1:3); pp. 173–183.
25. Wagner, M. J.; Gilman, P. (2011). Technical Manual for the SAM Physical Trough Model. TP-5500-51825. Golden, CO: National Renewable Energy Laboratory.
26. Wilcox, S.; Marion, W. (2008). Users Manual for TMY3 Data Sets. NREL/TP-581-43156. Golden, CO: National Renewable Energy Laboratory.
27. Wiser, R.; Bolinger, M. (August 2010). 2009 Wind Technologies Market Report. LBNL-3716E. Berkeley, CA: Lawrence Berkeley National Laboratory.

CHAPTER 5

FEASIBILITY OF GRID-CONNECTED SOLAR PV ENERGY SYSTEM: A CASE STUDY IN NIGERIA

MUYIWA S. ADARAMOLA

5.1 INTRODUCTION

Availability and utilization of energy is essential for social and economic development of a society and it is an essential resource required to improve human standard of living and quality of life. In Nigeria, access to reliable and stable supply of electricity is a major challenge for both the urban and rural dwellers. However, this problem is more significant in the rural areas and communities where only about 10% of the population have to access to electricity [1]. Even in the urban areas where grid-connected electricity is available, access to electricity is still a big challenge due to low and inadequate generation and distribution capacity. At the time of preparing this article, the peak electricity generation capacity in Nigeria is 3119.4 MW;

Reprinted from International Journal of Electrical Power & Energy Systems, *Volume 61, Adaramola MS, Viability of Grid-Connected Solar PV Energy System in Jos, Nigeria, pp. 64–69, Copyright 2014, with permission from Elsevier.*

which is about 24.4% of the peak electricity demand forecast of 12,800 MW for the same period [2]. This situation can be improved upon by using renewable energy resources, especially solar energy, to supplement the grid electricity supply in Nigeria. However, due to intermittent nature of these resources, they may not be suitable and reliable as stand-alone energy systems. Therefore, integration of both renewable energy conversion systems with storage facility could be a reliable energy system option in many locations in Nigeria.

In remote areas with no grid access, battery bank can be used as the storage facility. The negative effect of this is that additional cost of battery could significantly increase the unit cost of the electricity produced [3]. But, in areas with grid system, energy storage facility can be removed, and instead, the grid system can be used as 'storage' system. In this arrangement, when the renewable energy conversion system (RECS) produce more energy than needed, the surplus energy is fed into the grid and, energy is taken from the grid when the RECS system produces less energy than needed. As outlined by Mondal and Islam [4], some of the other advantages of PV-grid tied energy system includes: it can reduce energy and capacity losses in the utility distribution network, and it also can avoid or delay upgrades to the transmission and distribution network where the average daily output of the PV system corresponds with the utility's peak demand period. This arrangement is a good option for a PV-grid energy system in a tropical region like Nigeria, due to high availability of solar radiation in the country (see Section 2 below).

Feasibility, reliability and economic analyses conducted in a number of studies showed that hybrid power systems either as standalone or grid-tied system, are more reliable and cheaper than single source energy systems [see e.g., [5] and [6] and could produces less greenhouse gases when compare with fossil-fuel resources based energy systems [see e.g., [5], [6] and [7]. The objective of this work is to investigate the techno-economic viability of solar PV-grid connected energy system in a location north-east Nigeria. This energy system may not only improve access to reliable supply of electricity, but can also reduce dependency on diesel generator systems (which are commonly used to supplement grid supplied electricity in semi-urban and urban areas across the country), and thereby reducing the associated noise pollution and emissions from these diesel generators.

The energy system software, HOMER (Hybrid Optimization Model for Electric Renewable) is used to model the energy system and access its technical and economic performance.

5.5.2 SELECTED SITE AND SOLAR ENERGY RESOURCE

Nigeria is located in western Africa on the Gulf of Guinea and has a total area of 923,768 km^2 and lies between latitudes 4° and 14°N, and longitudes 2° and 15°E. The solar radiation distribution in Nigeria is shown in Fig. 1. Three distinct different solar radiation zones can be identified and they are labeled as I, II and III with each zone having different range of solar radiation. The solar radiation in each zone can roughly be group as:

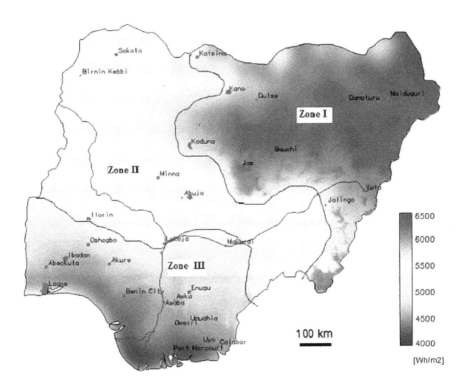

FIGURE 1: Solar radiation map of Nigeria [8] and [9].

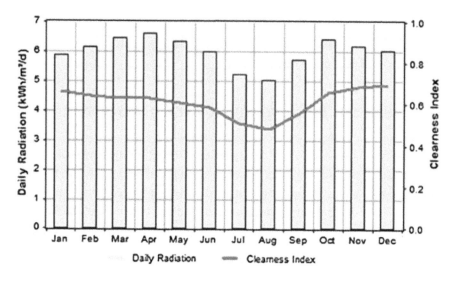

FIGURE 2: Monthly daily averaged global solar radiation and clearness index for Jos.

zone I: 5.7–6.5 kW/m²/day with sunshine hours of about 6 per day, zone II: 5.0–5.7 kW/m²/day with sunshine hours of about 5.5 per day and, zone III: 3.5–5 kW/m²/day with sunshine hours of about 5 per day.

For this study, a site (Jos, in Plateau state) located in the zone I is selected. This site is located on latitude 9°52′N and longitude 8°54′E and at an elevation of about 1238 m above sea level. The performance of a solar PV-grid-connected system is strongly dependent on the solar radiation which is site-specific and for a given site, its values vary frequently. Based on the data obtained from NASA Surface meteorology and solar energy website (NASA), the monthly averaged daily global solar radiation data for Jos is shown in Fig. 2. As expected, monthly variation in global solar radiation is observed and hence, the monthly energy output from solar PV would vary from one month to another. The monthly clearness index, which is defined as the fraction of solar radiation at the top of the atmosphere that reaches a particular location on the earth surface varied between 0.48 in the month of August (rainy season) and 0.70 in the month of December (dry season)

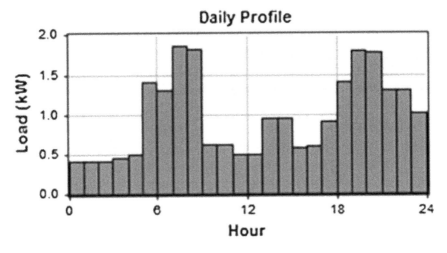

FIGURE 3: Typical daily electricity consumption in Jos Nigeria [5] and [10].

with an annual average of 0.61. The prevailing weather condition in Jos can generally be considered as partly overcast weather (but close to clear weather condition around November and December).

5.3 ELECTRICAL LOAD AND ENERGY SYSTEM COMPONENTS

The daily electrical load used in this study is taken from the work of Ogbonna et al. [10]. They reported in detail the domestic energy consumption patterns in Jos, northern Nigeria. From a typical daily electricity consumption profile for this location (see Fig. 3), two prominent peak demand periods can be observed in daily electricity load profile from this figure and they occur in the morning, between 06.00 and 09.00 and; late in the evening, between 19.00 and 21.00. These electrical load peaks are due to usual morning activities (e.g., cooking of breakfast, lighting), and cooking of supper, lighting, TV, reading (in the evening/night).

They also noted that there is an insignificant difference between the daily electricity demand patterns for weekdays and weekends and the daily average demand is reported to be about 1 kW h. However, the simulation analysis carried out in this study is based on assumed electrical load of 750 kW h/day for an area with reliable electricity grid access. At an average consumption rate of 2.5 kW h per day, 300 households will be benefit from this installation. For this load profile, hourly and daily variations are taken as 15% and 25% respectively.

The proposed energy system comprises of solar energy conversion system (PV) and national electricity grid system. The schematic diagram of the solar PV and grid-connected system is shown in Fig. 4. Detailed descriptions of each component with the required input data are presented in the following sections.

5.3.1 GRID SYSTEM

The grid purchase capacity and sales are taken as 100 kW and 60 kW, respectively and interconnection charge is assumed to be $500. In this study,

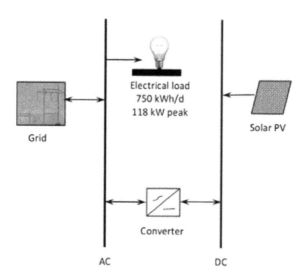

FIGURE 4: The schematic diagram of the solar PV and grid-connected system.

it was assumed that the consumers will buy certain amount of electricity from the grid at a price specified by the distribution company. The excess electricity generated will be sold back to the grid at 75% of grid price. In Nigeria, the cost of electricity is fixed and regulated by the Nigerian Electricity Regulatory Commission (NERC). The monthly electricity bill is made-up of fixed charge and energy consumed charge. Both of these charges vary across the country. The current monthly (2014) energy charge and fixed charge by distribution company in Jos area for residential users (R2) are N14.32/kW h and N1163, respectively and for residential users (R3), energy charge and fixed charge are N25.80/kW h and N43625, respectively [5]. The R2 and R3 users generally consumed more than 50 kW h of electricity per month and have the voltage connection systems of single and 3-phase, and LV (Low Voltage) Maximum demand, respectively.

TABLE 1: Technical specification of MaxPower CS6X-300P under nominal operating cell temperature [13].

Parameter		Specification
Nominal maximum power		300 Wp
Optimum operating voltage		32.9 V
Optimum operating current		6.61 A
Open circuit voltage		41.0 V
Short circuit current		7.19 A
Efficiency		14.71%
Operating temperature		−40 °C ~ +85 °C
Temperature coefficients:	P_{max}	−0.43%/°C
	V_{oc}	−0.34%/°C
	I_{sc}	0.065%/°C
Normal operating cell temperature		45 ± 2 °C
Dimension		1954 mm × 982 mm × 40 mm

5.3.2 PV ARRAY AND CONVERTER

The solar PV-grid system consists of 80 kW PV module. The solar PV module is a 72-cell (6 × 12) poly-crystalline (model number CS6X-300M, manufactured by Canadian Solar) with a rated power of 300 Wp and can produced maximum 600 V DC. The detailed technical specifications of this model are presented in Table 1. The lifetime of the PV array is assumed to be 25 years (which is equivalent to worldwide warranty provided by the manufacturer). The initial cost of the PV is taken as US$2322 (or US$2400) per kW (Adaramola et al. [5]). The derating factor and the ground reflectance are taken as 80% and 20%, respectively and no tracking system is included in the PV system.

Since the power output from the solar PV module is in DC, power inverter system is required to converter the PV power output to AC power. The size of the converter is determined using Solectria String Sizer tool [11] and it was found that one PVI 85KW/PVI 85KW-PE inverter model can supply the required AC power output. This inverter model is designed for grid-tied electricity system. The technical specifications of this model are presented in Table 2. The cost of this inverter model is given as $28,250 [12]. The shipment, import duty and related costs are assumed to be 30% of the cost of the inverter. Therefore, the total initial cost of the inverter model is taken as US$36725 (or US$432/kW). The lifetime of a unit is taken as 15 years with an average efficiency of 90%.

TABLE 2: The technical specifications of the selected inverter [14].

Parameter		Specifications
DC Input	Absolute max voltage	600 VDC
	MMPT voltage range	300–500 VDC
	Max operating	264 A
AC Output	Nominal voltage	208, 240, 480 or 600 VAC, 3-Ph (3 wire standard, 4 wire option)
	Continuous power (VAC)	85 kW
	Continuous current (VAC)	236/205/102/82A
Others	Peak efficiency	96.6/96.5/97.0/96.9/97.5%
	Ambient temp. range (full power)	−40°F to +131°F

5.3.3 PROJECT LIFE AND ECONOMY

The lifetime of the project is taken to be 25 years. According to the available information, the current interest rate and inflation rate in Nigeria are 12% and 8% respectively [15] and from this these, the annual real interest rate is determined as 3.7% using Fisher expression. This value represents current real interest rate. The initial fixed capital cost which can be used to prepare the site for the system and other initial installation costs is taken as $25000 and the overall system operation and maintenance cost of $1500 per year is assumed.

5.3.4 HOMER SOFTWARE

The energy system is designed and analyzed using HOMER software. Among the available softwares, HOMER is the most widely used optimization software for hybrid systems [16] and [17]. This is due to its many possible combination of renewable energy systems and ability to perform optimization and sensitivity analysis which makes it easier and faster to evaluate the many possible system configurations [16]. There are many studies that have used HOMER to examine the technical and economic feasibility of hybrid energy systems worldwide [4], [5], [6], [7], [18], [19], [20], [21], [22] and [23]. Detailed description of this software can be found in [24].

The output power of a PV array can be calculated from the following equation and the PV specifications [23].

$$P_{PV} = Y_{PV} f_{PV} \left(\frac{\overline{G_T}}{G_{T,STC}} \right) \left[1 + \alpha_p \left(T_c - T_{c,STC} \right) \right]$$

(1)

where P_{PV} is the rated capacity of the PV array, that is, the power output under standard test conditions in kW; f_{PV} is the PV derating factor [%], G_T is the solar radiation incident on the PV array in the current time step [kW/m²], $(G_{T,STC})$ is the incident radiation at standard test conditions [1 kW/

m²], α_p is the temperature coefficient of power [%/°C], T_c is the PV cell temperature in the current time step [°C] and $T_{c,STC}$ is the PV cell temperature under standard test conditions [25 °C]. In a case where the effect of temperature on the PV array performance is neglected, α_p can assumed to be zero and Eq. (1) reduces to:

$$P_{PV} = Y_{PV} f_{PV} \left(\frac{\overline{G_T}}{G_{T,STC}} \right)$$

(2)

5.4 RESULTS AND DISCUSSION

5.4.1 ELECTRICITY GENERATION

The electricity generated by the PV-grid energy system and the corresponding electricity consumed by the users (when the cost of PV is $2400/kW and global solar radiation is 6 kW h/m²/day) is shown in Table 3. This table shows the total electricity produced by the energy system is 331,536 kW h/year which comprises of 133,867 kW h/year (40.4%) from the solar PV and 197,668 kW h/year (59.6%) from the grid. The generated electricity is utilized as follows: 82.6% (273,695 kW h/year) is consumed by the users, 13.4% (44,454 kW h/year) is sold back to the grid and 4.0% accounted for transmission losses.

TABLE 3: Electricity generated by the solar PV-grid system and end-use consumption pattern.

	Production			Consumption	
	kW h/yr	%		kW h/yr	%
Solar PV	133,867	40.4	AC load	273,695	82.6
Grid purchases	197,669	59.6	Grid sales	44,454	13.4
			Losses	13,387	4.0
Total	331,536	100.0	Total	331,536	100.0

The performance parameters of the solar PV are presented in Table 4. The capacity factor, which is defined as the ratio of an actual energy output during a given period to the energy output that would have been generated if the system is operated at full capacity for the entire period, is given as 19.1%. This relatively low value is due to facts that in this study, the orientation of the system fixed tilt (at latitude). The capacity factor can be increased using tracking system. This factor is a useful parameter for users, developer and manufacturer of the solar PV energy system. It determines the economic viability of this energy system. The value of capacity factor for the PV determined in this study is comparable to findings from similar studies (see e.g., [25]). It can further be observed that the average daily energy output from the solar PV system is observed to be 367 kW h/day or 15.29 kW.

TABLE 4: Performance parameters for the PV system.

Parameters	Quantity
Rated capacity (kW)	80
Mean output (kW)	15.3
Mean output (kW h/day)	367
Capacity factor (%)	19.1
Total production (kW h/year)	133,867

The monthly average of electricity produced by each component of the PV-connected energy system is presented in Fig. 5. The monthly and seasonal variations in amount of electricity produced by PV system and contribution from the grid can be observed from this figure. This is due to the variability in monthly global solar radiation. The maximum average monthly power generated by the solar PV is about 16.9 kW (in November) and the minimum of about 12.2 kW occurred in August. The monthly energy from purchased and sold to the grid is presented in Table 5. It can be observed from this table that the quantity of electricity purchased varies from 14,941 kW h in November when the PV produced highest amount electricity to 19,486 kW h in August when the PV produced the least amount of electricity. Similar variations in the electricity sold to the grid can also be observed from this table.

TABLE 5: Monthly amount of electricity purchased from and sold to the grid.

Month	Energy purchased (kW h)	Energy sold (kW h)
January	16,577	4503
February	14,089	4163
March	17,210	4154
April	16,104	3844
May	15,658	3645
June	16,914	3036
July	17,261	2480
August	19,486	2356
September	16,823	3202
October	16,327	4227
November	14,914	4404
December	16,306	4440
Annual	197,669	44454

5.4.2 ECONOMIC ANALYSIS

The economic analysis of the solar PV-grid connected system is assessed by the following indicators: the levelized cost of electricity (LCOE) and net present cost (NPC) of the system. The effects of the cost of the PV and global solar radiation on these indicators are also presented in this section. The HOMER software determine the net present cost of the system as the different between the present value of all the costs of installing and operating the system over its project lifetime, and the present value of all the revenues that it earns over the project lifetime. The variation of the NPC and LCOE with unit cost of the PV array is presented in Table 6. It should be mentioned that the presented LCOE and NPC are based on the net metering on monthly purchases from grid. For the base case; PV cost of $2400/ kW, global solar radiation of 6.0 kW h/m²/day and grid price of electric-

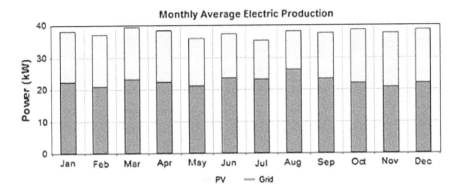

FIGURE 5: Monthly distribution of the electricity produced by the energy system.

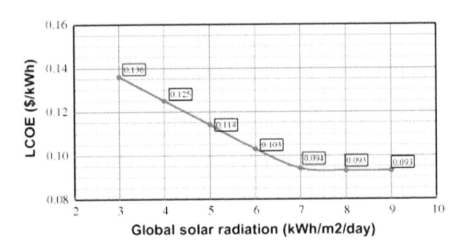

FIGURE 6: Effect of global solar radiation on the LCOE when the PV cost is $2400/kW.

ity of \$0.090/kW h (for R2 residential users), the NPC for this project is \$530,090, annual operating cost is \$17,103 and the cost of the electricity is \$0.103/kW h. As the cost of the PV increases, the LCOE, as expected, increases gradually from \$0.103/kW h (when the PV is cost of \$2400/kW) to \$0.159/kW h (when the PV is cost of \$6000/kW). The observed increment in the LCOE is primarily due to the increase in the initial capital cost of the PV which resulted in higher NPC when compared with the base case. It is expected that at lower PV cost (compared with the base case), the LCOE of the system will be reduced. For instance, when the simulation analysis was performed for PV cost of \$1800 and \$1200 per kW, the LCOE is found respectively as \$0.081/kW h and \$0.065/kW h. This shows that for the location considered in this study and similar locations in Zone I on the solar radiation map of Nigeria (see Fig. 1), solar PV-tied integration energy system is feasible and economically viable energy system.

The effect of global solar radiation on the LCOE is shown in Fig. 6 when the PV cost is \$2400/kW. Two regions can be observed in this figure. These regions are: region 1 (when global solar radiation is less than or equal to about 7.0 kW h/m^2/day)—a linear relationship exists between the global solar radiation and the LCOE. In this region, the LCOE is observed to decrease with increasing global solar radiation. In region 2 (when global solar radiation is greater than about 7.0 kW h/m^2/day), the LCOE (at \$0.093/kW h) is observed to be constant irrespective of the global solar radiation.

TABLE 6: Effect of PV cost on the LCOE and NPC.

PV Cost (/\$/kW)	Initial capital (\$)	Operating cost (\$/yr)	Total NPC (\$)	LCOE (\$/kW h)
2400	254,220	17,103	530,090	0.103
3000	302,220	17,103	578,090	0.113
3600	350,220	17,103	626,090	0.122
4200	398,220	17,103	674,090	0.131
4800	446,220	17,103	722,090	0.141
6000	542,220	17,103	818,090	0.159

5.5 CONCLUSIONS

This study examines the feasibility of solar PV-grid tied energy system for electricity generation in a selected location in the northern part of Nigeria. The technical and economic performance of a combination of 80 kW solar PV and 100 kW power from the grid was investigated. It was found that this energy system can generates annual electricity of 331,536 kW h with solar PV contributing 40.4% and the levelized cost of energy is found to $0.103/kW h.

It is further observed that by reducing the initial installation costs (which consists of capital cost of the PV, connection cost and other associated costs), the cost of electricity can be significantly reduced. In addition, it was observed that the global solar radiation plays significantly impact on the economic viability of this system. It is expected that incorporating solar PV with grid system can reduce carbon dioxide and other pollutant emissions associated with thermal power plants generated electricity.

The information presented in this paper can serve as input to the development of grid-connected solar PV energy system in Nigeria. Based on the findings from this study, the development of grid-connected solar PV system in the north-eastern part of Nigeria could be economically viable energy system. However, as a result of high initial investment cost of solar PV system, favorable policies and incentives from government can accelerate the development of this type energy system in Nigeria. The logical next step from this study should be installation and performance assessment of practical grid-connected solar PV system in this location.

REFERENCES

1. M.S. Adaramola, O.M. Oyewola, S.S. Paul. Technical and economic assessment of hybrid energy systems in South-West Nigeria. Energy Explor Exploit, 30 (4) (2012), pp. 533–552
2. Nigeria-power-reform. [assessed 22.01.2014]. <http://nigeriapowerreform.org>.
3. H. Häberlin. Photovoltaics system design and practice. (1st ed.)John Wiley & Sons Ltd, Chichester (2012)
4. M.A.H. Mondal, A.K.M.S. Islam. Potential and viability of grid-connected solar PV system in Bangladesh. Renew Energy, 36 (2011), pp. 1869–1874

5. M.S. Adaramola, S.S. Paul, O.M. Oyewola. Assessment of decentralized hybrid PV solar-diesel power system for applications in Northern part of Nigeria. Energy Sustain Develop, 19 (2014), pp. 72–82

6. J. Tang, B. Ye, Q. Lu, D. Wang, J. Li. Economic analysis of photovoltaic electricity supply for an electric vehicle fleet in Shenzhen, China. Int J Sustain Transport, 8 (3) (2014), pp. 202–224

7. A. Hossam-Eldin, A.M. El-Nashar, A. Ismaiel. Investigation into economical desalination using optimized hybrid renewable energy system. Int J Electr Power Energy Syst, 43 (1) (2012), pp. 1393–1400-

8. Huld T, Šúri M, Dunlop E, Albuisson M, Wald L. Integration of HelioClim-1 database into PVGIS to estimate solar electricity potential in Africa. In: Proceedings from 20th European photovoltaic solar energy conference and exhibition, Barcelona, Spain; 2005. [accessed 29.03.13]. <http://re.jrc.ec.europa.eu/pvgis/>.

9. O.S. Ohunakin, M.S. Adaramola, O.M. Oyewola, R.O. Fagbenle. Solar energy applications and development in Nigeria: drivers and barriers. Renew Sustain Energy Rev, 32 (2014), pp. 294–301

10. A.C. Ogbonna, O. Onazi, J.S. Dantong. Domestic energy consumption patterns in a Sub-Saharan. African City: the study of Jos-Nigeria. J Environ Sci Resource Manage, 3 (2011), pp. 48–62

11. http://www.solren.com/products-and-services/string-sizing-tool/.

12. http://www.affordable-solar.com/store/solar-inverters-commercial.

13. Canada-solar. <www.canadiansolar.com>. [accessed on 15.01.2013].

14. http://www.civicsolar.com/. [accessed on 15.01.2013].

15. Central Bank of Nigeria. [accessed on 21.01.2014].

16. J.L. Bernal-Agustin, R. Dufo-Lopez. Simulation and optimization of stand-alone hybrid renewable energy systems. Renew Sustain Energy Rev, 13 (2009), pp. 2111–2118

17. S. Sinha, S.S. Chandel. Review of software tools for hybrid renewable energy systems. Renew Sustain Energy Rev, 32 (2014), pp. 192–205

18. J. Dekker, M. Nthontho, S. Chowdhury, S.P. Chowdhury. Economic analysis of PV/diesel hybrid power systems in different climatic zones of South Africa. Int J Electr Power Energy Syst, 40 (1) (2012), pp. 104–112

19. J.B. Fulzele, S. Dutt. Optimum planning of hybrid renewable energy system using HOMER. Int J Electr Comput Eng, 2 (1) (2012), pp. 68–74

20. P.M. Murphy, S. Twaha, I.S. Murphy. Analysis of the cost of reliable electricity: a new method for analyzing grid connected solar, diesel and hybrid distributed electricity systems considering an unreliable electric grid, with examples in Uganda. Energy, 66 (2014), pp. 523–534

21. M. Rohani, G. Nour. Techno-economical analysis of stand-alone hybrid renewable power system for Ras Musherib in United Arab Emirates. Energy, 64 (2014), pp. 828–841

22. Sureshkumar U, Manoharan PS, Ramalakshmi APS. Economic cost analysis of hybrid renewable energy system using HOMER. In: Proceedings of the IEEE – international conference on advances in engineering science and management, 30–31 March; 2012. p. 94–99.

23. S.M. Shaahid, L.M. Al-Hadhrami, M.K. Rahman. Review of economic assessment of hybrid photovoltaic-diesel-battery power systems for residential loads for different provinces of Saudi Arabia. Renew Sustain Energy Rev, 31 (2014), pp. 174–181

24. T. Lambert, P. Gilman, P. Lilenthal. Micropower system modeling with HOMER. F.A. Farret, M.G. Simões (Eds.), Integration of alternative sources of energy, John Wiley and Sons (2006)

25. Ayompe LM, Duffy A. An assessment of the energy generation potential of photovoltaic systems in Cameroon using satellite-derived solar radiation datasets. Sustainable Energy Technologies and Assessments; 2013. <http://dx.doi.org/10.1016/j.seta.2013.10.002>.

CHAPTER 6

PASSIVE COOLING TECHNOLOGY FOR PHOTOVOLTAIC PANELS FOR DOMESTIC HOUSES

SHENYI WU AND CHENGUANG XIONG

6.1 INTRODUCTION

The operating temperature is one of the important factors that can affect the efficiency of the PV panels. The effects of temperature on photovoltaic efficiency can attribute to the influences on the current and voltage of the PV panels. This can be easily found on the I-V curve of the panels. It results in a linear reduction in the efficiency of power generation as temperature increases [1]. The efficiency of some types of PV cells is very much dependent on their operating temperature. For crystalline silicon solar cells, the reduction in conversion efficiency is 0.4–0.5% for every degree of temperature rise [2]. Therefore, reducing the operating temperature of photovoltaic cells is important for the PV panel to work efficiently and protect cells from irreversible damage.

Passive Cooling Technology for Photovoltaic Panels for Domestic Houses. © *Wu S and Xiong C.* International Journal of Low-Carbon Technologies *0 (2014). doi:10.1093/ijlct/ctu013. Licensed under the Creative Commons Attribution 3.0 Unported License, http://creativecommons.org/licenses/by/3.0/.*

A number of researchers have worked on cooling the PV panels with different approaches. Air circulation is probably the most simple and natural way for this purpose. In order to enhance convection heat transfer, fins were used to extend the heat transfer area. Edenburn [3] developed a device, made up of linear fins on all available heat sink surfaces, used for cooling single cells passively. Araki et al. [4] did a further research on passive cooling technologies and found that good thermal conduction between cells and heat spreading plate was important. Combining PV and solar thermal collectors (PV/T) is another way of cooling PV panels. Tonui and Tripanagnostopoulos [5] reported their experiment on modified PV/T collectors, and results showed the maximum temperature reduction achieves 10°C by natural ventilation and 308C by forced ventilation.

As a good cooling media, water has been widely used for PV cooling in various forms. It is very suitable for PV/T systems. Kalogirou [6] studied a water-based PV/T system consisting of four monocrystalline PV panels in the Cyprus and achieved an increase of average annual electrical efficiency from 2.8 to 7.7% with the payback periods of 4.6 years. Tripanagnostopoulos et al. [7] compared electrical efficiency of PV/WATER, PV/AIR and PV/FREE and PV/INSUL under ambient air temperature of 29°C. They achieved the maximum increase by 3.2% with PV/ WATER.

Krauter [8] investigated the method of covering PV modules with a flowing water film above.With the additional evaporation heat transfer, it was claimed that they could decrease the cell temperature up to 22°C and obtained a net increase from 8 to 9%. Abdolzadeh and Ameri [9] used water spray to cool the PV panels and achieved increasing the efficiency of cells by 3.26 to 12.5%. Kordzadeh [10] studied that a thin continuous film of water running on the front of the surface of modules obtained better electrical efficiency because of reducing reflection loss and surface temperature.

To avoid additional energy consumption incurred for cooling the PV panels, Furushima and Nawata [11] reported a model with cooling water being supplied from a city water supply system by Siphonage and the cooling system did not require any additional energy input on the site. Wilson [12] studied the gravity-fed technology where water was transported from upstream sources like river to downstream sources by gravity. The results obtained from this work showed a 12.8% increase in electrical efficiency as a result of 32°C temperature reduction.

Other technologies were also used to enhance the heat transfer for cooling the panels. Akbarzadeh and Wadowski [13] reported an innovative gravity-assisted heat pipe system to optimize the cooling of concentrated photovoltaics. It was found that the temperature at the surface of solar cells did not exceed 46°C during a 4-h test, and the efficiency was increased by 50%. Huang et al. [14] initially integrated PCM into BIPV system and used fins for improvement. Biwole et al. [15] established a numerical model and used CFD to simulate heat and mass transfer of PCM at the back of photovoltaic panels. Their results showed that adding PCM at the back of panels can maintain the operating temperature below 40°C.

Active cooling is effective to cool PV panels. However, with the additional power consumption involved, the active cooling purely used to lower the operating temperature does not have obvious benefit in the net gain of efficiency. The technologies such as PV/T (photovoltaic thermal) system or the PV-SAHP (photovoltaic solar heat pump) system [16, 17] seem to address the issue stated earlier by combination of two systems. But the fact that PV/T has to at a higher operating temperature in order to supply useful heat means the gain by cooling is limited. What is more, the higher initial investment and the final benefit with PV/T technology is contributed to thermal energy rather than electricity [7]. This renders the PV/T being not an effective technology for the original purpose. Therefore, finding a simple and feasible way to cool the PV panel without requiring further energy input is still much sought after.

6.2 DESCRIPTION OF THE SYSTEM

Figure 1 illustrates an example of the proposed solar-driven rainwater cooling system. The system consists of a PV module with an area of 1.46 m², maximum efficiency of 15.4% and maximum power output of 250 W, a gas expansion chamber, a rainwater storage tank and a secondary water tank. A cylindrical gas expansion chamber is installed at the eaves whereas the secondary water tank, which is connected to the gas expansion chamber, is hung at the side of the house. Gutters are installed on both south- and north-facing roofs in order to maximize the rainwater harvesting. On receiving the solar radiation, the gas in the chamber expands with the temperature increase. The rainwater in the tank is pushed upwards by the expanding gas so that it flows over the PV panel through a distribution tube on the top as shown in Figure 2. The rainwater is not considered being reused to reduce the cost and simplify the system structure in this case.

6.3 ANALYSIS OF THE SYSTEM

The amount of the rainwater delivered to the PV panels is determined by the gas expansion volume. The expansion volume is a function of temperature that varies with the solar energy the gas received. Therefore, there is a relationship between the amount of rainwater delivered and solar incidence. The relationship can be derived from the energy conservation law.

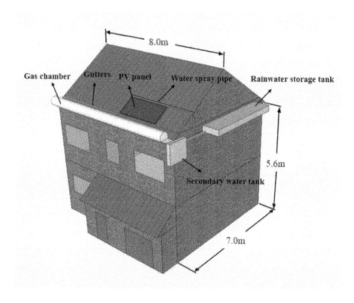

FIGURE 1: 3-D model of the solar-driven rainwater cooling system installed on the roof.

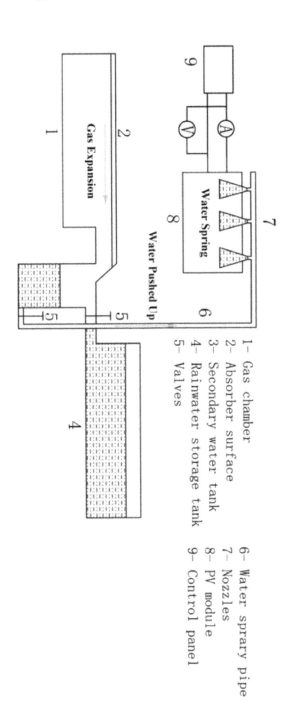

FIGURE 2: Schematic diagram of cross section of the solar-driven rainwater cooling system.

1– Gas chamber
2– Absorber surface
3– Secondary water tank
4– Rainwater storage tank
5– Valves
6– Water sprary pipe
7– Nozzles
8– PV module
9– Control panel

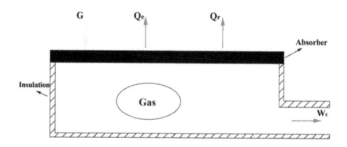

FIGURE 3: Schematic diagram of cross section of the gas chamber.

6.3.1 ENERGY BALANCE EQUATION IN THE GAS EXPANSION CHAMBER

The gas expansion chamber is covered by an insulation layer used to reduce heat loss from the side surfaces and is covered with an absorption layer to enhance the capture of solar radiation. Taking the chamber as a control volume, the energy conversion and heat flows of the chamber are as shown in Figure 3. The energy balance can be expressed as $Q_{net} = G - Q_c - Q_r - W_E$; where W_E is gas expansion work (J). The Q_{net} is the heat that causes the gas temperature rise.

To simplify the simulation, some assumptions are made and stated as follows:

- No heat transfer across the side boundaries. The conduction heat transfer between the absorption layer and the surrounding air is neglected.
- Thermal resistances in the absorption layer and gas are not considered so that the absorption layer and gas have the same temperature and the gas temperature in the chamber is uniform.
- The gas is treated as ideal gas.
- The atmospheric pressure is constant within duration of time.
- The gas properties are constant.

Under the assumptions, the energy balance of the gas expansion chamber can be expressed in the following equation:

$$A_{ab} \propto \int I\, dt - A_{ab}h_c \int (T_{gas} - T_a)\, dt$$
$$-A_{ab}\varepsilon\sigma \int (T_{gas}^4 - T_a^4)\, dt - P_{atm}\delta_V$$
$$= m_{gas}C_{gas}\Delta T_{gas} \tag{1}$$

where I is solar radiation on horizontal surface (W/m²); T_{gas} represents gas temperature (°C); A_{ab} represents area of absorber surface (m²); C_{gas} denotes specific heat capacity of gas (J/kg K); h_c denotes convection heat transfer coefficient of wind (W/m² K); α is absorption coefficient of PV panel (0.95); ε is emissivity factor of absorber (0.04); σ is Stefan–Boltzmann constant (W/m² K⁴); t represents time (s).

This equation describes the accumulating effect of the solar radiation on the gas expansion chamber from a reference point and its derivative form:

$$A_{ab} \propto I - A_{ab}h_c(T_{gas} - T_a) - A_{ab}\varepsilon\sigma(T_{gas}^4 - T_a^4)$$
$$-P_{atm}\frac{dV}{dt} - m_{gas}C_{gas}\frac{dT}{dt} = 0 \tag{2}$$

It describes the effect of the solar radiation on the gas chamber at any time point. Since the chamber's temperature change is a slow process, we use the finite-difference equation to approximate Equation (2) as follows:

$$A_{ab} \propto I - A_{ab}h_c(T_{gas} - T_a) - A_{ab}\varepsilon\sigma(T_{gas}^4 - T_a^4)$$
$$-\frac{P_{atm}(A_w H_{(t)} - A_w H_{(t-1)})}{\Delta t}$$
$$= \frac{m_{gas}C_{gas}(T_{gas,(t)} - T_{gas,(t-1)})}{\Delta t} \tag{3}$$

where the subscribes (t) and (t − 1) denote the time step in hour and AW is the base area of the secondary tank (m²) and H is the height of water pumped (Figure 4).

For 1-h time interval, we have the following equation:

$$A_{ab} \propto I - A_{ab}h_c\left(T_{gas} - T_a\right) - A_{ab}\varepsilon\sigma\left(T_{gas}^4 - T_a^4\right)$$

$$-\frac{P_{atm}\left(A_wH_{(t)} - A_wH_{(t-1)}\right)}{3600}$$

$$= \frac{m_{gas}C_{gas}\left(T_{gas,(t)} - T_{gas,(t-1)}\right)}{3600} \tag{4}$$

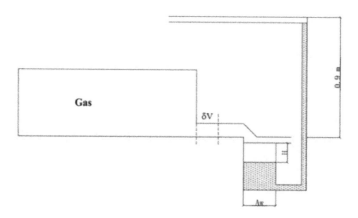

FIGURE 4: Schematic diagram of cross section of the gas chamber and secondary water tank.

6.3.2 HYDRAULIC HEAD OF WATER IN THE SECONDARY WATER TANK

With reference to Figure 4, the expression of hydraulic head in the secondary water tank varying with the gas expansion can be derived as follows.

At initial state, the secondary water tank is filled with water and the gas in the chamber does not expand. The state of the gas can be expressed as follows:

$$P_{atm}V_{gas} = m_{gas}RT_0 \qquad (5)$$

where T_0 denotes initial temperature of the chamber (°C); P_{atm} indicates atmosphere pressure (Pa).

On receiving heat from the solar radiation, the gas in the chamber starts to expand. If the gas volume is expanded by δV (volume expansion) (m³), the same volume of water will be pushed out of the tank. The change of the state follows the following equation:

$$[P_{atm} + \rho g(0.9 + H)](V_{gas} + \delta_v) = m_{gas}RT_{gas} \qquad (6)$$

where m_{gas} represents mass of gas (kg); V_{gas} denotes volume of gas (m³) and $\delta_v = H \times A_w$; where AW = 0.25: From Equations (5) and (6), a relationship between the volume of the water pushed and gas temperature can be obtained:

$$\rho g A_W H^2 + \left(P_{atm}A_W + \rho g V_{gas} + 0.9\rho g A_W\right)H + P_{atm}V_{gas}$$

$$+0.9\rho g V_{gas} - \frac{P_{atm}V_{gas}T_{gas}}{T_0} = 0 \qquad (7)$$

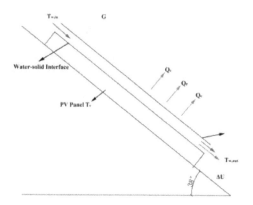

FIGURE 5: Schematic diagram of cross section of the PV panel.

6.3.3 HEAT TRANSFER ON THE PV PANEL

When the water flows over a tilted PV panel, the heat transfer between the water and panels can be complicated by involving water evaporation in addition to the normal radiation and convection heat transfer. With reference to Figure 5, the energy balance leads to:

$$G = Q_e + Q_c + Q_r + \Delta U$$

where G denotes energy generated from solar radiation (J), U indicates internal energy (J), and Q_e, Q_c, and Q_r denote heat loss by evaporation (J), heat loss by convection (J) and heat loss by radiation (J), respectively.

The heat transfer mechanism is quite complex due to temperature variation along the water–solid interface. A two-dimensional steady-state model is used, and some assumptions should be made to simplify the calculation.

- Solar radiation irradiates on the PV panel, 15.4% is converted to electricity energy, 5% is reflected and the rest part is converted to heat energy.
- Assume the water mass is uniformly distributed over the PV panel and water is ultimately heated to a temperature that is same as the cell temperature T_c.
- Water temperature increase caused by solar radiation is neglected. [18]
- Convective heat loss at the back of PV panel and radiation heat transfer is not considered.

Air flowing at the air–water interface essentially accelerates water evaporation rate. Thus, the air convection could be measured accompanying with water evaporation. Smith et al. [19] predicted evaporation heat transfer flux by the following equation, which approximately estimates how much latent heat is removed from the PV panel by water evaporation (w/m²).

$$q_e = (0.0638 + 0.0669V)(P_w - P_a) \qquad (8)$$

where saturation pressure of water is as follows:

$$\log\left(\frac{P_s}{1000}\right) = 30.59051 - 8.2\log(T_a + 273.16)$$

$$+0.00248(T_a + 273) - \frac{3142.31}{(T_a + 273)}$$

partial pressure of water vapour at surrounding air is as follows:

$$P_a = \varphi P_s$$

partial pressure of water vapour at water is as follows:

$$P_w = \exp\left[20.386 - \frac{5132}{0.5(T_{w,in} + T_c)}\right] \times 133$$

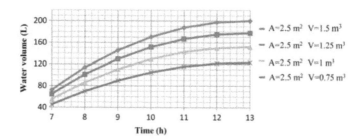

FIGURE 6: Accumulated rainwater volume pushed by gas chambers with different sizes.

where $T_{w,in}$ denotes water inlet temperature (8C).

Therefore, total heat transfer via water evaporation can be expressed as follows:

$$Q_e = A_{cell} q_e \tag{9}$$

where A_{cell} denotes area of cells (m²).

The internal energy change of water can be expressed as follows:

$$\Delta U = \frac{m_{water} C_p (T_{w,out} - T_{w,in})}{3600} = \frac{m_{water} C_p (T_c - T_{w,in})}{3600} \tag{10}$$

where $T_{w,out}$ denotes water outlet temperature (°C).

Thus, after a water film flowing down to a PV panel, the cell temperature can be approximately calculated as follows:

$$\alpha I A_{cell}(1 - \eta) = A_{cell}(0.0638 + 0.0669 U_W (P_w - P_a)$$

$$+ \frac{m_{water} C_a (T_c - T_{w,in})}{3600} \tag{11}$$

where U_w denotes wind speed (m/s).

6.4 SIMULATION RESULTS

This system was analysed with the climate data on a clear day of 29th July in Nottingham [20]. The day was chosen for its low wind speed and high air temperature. The study was focused on the thermal performance of the gas chamber and the cooling effect to the PV panels. The analysis was on one PV module system.

6.4.1 INFLUENCE OF DESIGN PARAMETERS ON THE SYSTEM PERFORMANCE

The size of the gas chamber is a predominant parameter that influences how much water can be pushed out of a tank. To evaluate the influences of chamber surface area and chamber volume on performance, two groups of

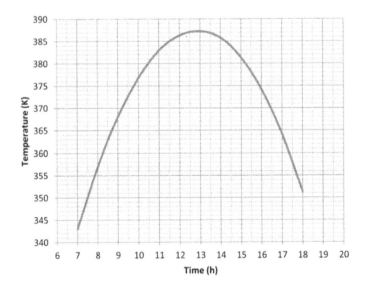

FIGURE 7: Gas temperature variation in a design day.

the gas chambers were analysed. In the first group, the chambers have the same surface area of 2.5 m² but with different volumes of 1.5, 1.25, 1 and 0.75 m³, respectively. The second group has a fixed volume of 1.25 m³ but with the different surface areas of 3, 2.5, 2 and 1.5 m², respectively. In this system, water is pushed up through a 0.9-m-high vertical pipe (Figure 4).

The results show that, for the same chamber volume, the gas temperature in the chamber slightly increases with the surface area. But the increase is within 1°C. This could be a result of the heat gain from the larger surface area being offset by the heat losses from the same larger surface area. Figure 6 presents accumulated water volume pushed due to gas expansion. It can be seen that the amount of water pumped increases with the gas chamber volume. It increases from 123 l/day with a 0.75-m³ chamber to 200 l/day with a 1.5-m³ gas chamber. The amount pumped increasing with the chamber volume is due to the assumption of uniform air temperature in the chamber. The volume expands more with the air mass increases in the chamber. However, this phenomenon should become less significant when the air temperature profile in the chamber is treated as non-uniform. Without any control, the gas chamber can pump the maximum amount of water to the PV panel at 7 am, and the amount gradually decreases to zero around 1 pm. It was estimated approximately that 165 l/day of rainwater is available for the climate under the consideration. To pump this amount of water, a 1-m³ gas chamber is needed.

6.4.2 PERFORMANCE OF THE PV MODULE COOLED BY THE SYSTEM

The performance is evaluated from the rainwater pumped by the device with a 0.16-m³ secondary water tank connecting to the a 1-m³ gas chamber. The secondary water tank is designed to protect gas infiltration and increase air tightness so that the whole system can work with the higher efficiency. The variation of the gas temperature in the chamber during a day was illustrated in Figure 7. The initial gas temperature is 293 K, which is quickly heated to 342 K at 7 am, because low heat capacity of gas, small mass of gas in the chamber and intensive solar radiation

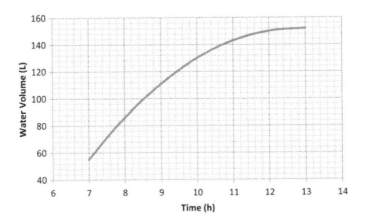

FIGURE 8: Accumulated rainwater volume in each hour.

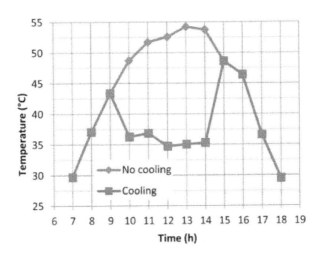

FIGURE 9: Comparison of operating temperatures between with cooling and without cooling to the PV panel.

on the design day cause that gas can be heated to a high temperature. With the increase of solar radiation, the gas temperature gradually increases to a maximum value 387 K at 1 pm. After that, gas temperature reduces due to the reduction of solar radiation and the rise of heat loss from a gas chamber to the outside environment.

TABLE 1: Climate data of the day of calculation.

Time	Temperature (^{0}C)	Wind speed (km/h)	Humidity (%)	Beam solar radiation (w/m^2)	Diffuse solar radiation (w/m^2)
7	16.3	6	80	517	122
8	18.2	4.6	72	528	207
9	20.3	4.3	67	634	193
10	22.3	4.3	62	744	166
11	24	4.3	59	712	232
12	25.2	6.5	56	682	249
13	26.2	6.5	54	798	149
14	27	5.4	52	768	153
15	27	4.3	51	558	216
16	26.7	7	53	646	128
17	25.6	10.3	57	381	134
18	24.2	6.5	61	463	59

TABLE 2. Hourly rainwater supply with temperature sensitive control for cooling.

Time	Water volume (l)
10	28.37
11	30.13
12	30.62
13	31.57
14	31.31

Under this condition, the device is able to push 152 l of water to the PV panel (Figure 8). As discussed earlier, without control, the majority amount of the water is pumped at the early time of the day when the demanding for cooling is low. A control to the flow may be needed, for example, by a temperature sensitive valve to delay the water pumping to address this issue. The operating temperature of PV is primarily determined by the solar radiation. On the day of 29th July, between 10 am and 14 pm, solar radiation was .850 W/m2 and its temperature reached 50°C. The detail of the climate data for the day is shown in Table 1. In order to maximize the cooling benefit to the PV panel, a temperature sensitive valve can be used to adjust the flow rate of water according to the roof temperature. Table 2 shows that a total of 152 l of water can be pushed at different hourly rates with respect to the roof temperature from 10 am to 2 pm on the day. It can be seen that with the temperature sensitive valve, more water is pumped when the roof temperature is higher at late hours, which allows more cooling to the PV penal when it receives high solar radiation.

During the working time, the cooling to the PV panel is very effective when the PV panel temperature is high as shown in Figure 9. It can be seen that at 1 pm, a maximum temperature reduction of 19°C is achieved and at other time temperature reduction ranges from 12.5 to 18.5°C. Figures 10 and 11 present the efficiency and the power output of the PV panel with and without cooling, respectively. The cooling maintains the efficiency of the cells above 14.5% each hour in a design day, particularly, between 12 pm and 2 pm during which the PV panel has very low efficiency without cooling. The cooling also increases the power output by 16W on average. In summary, solar-driven cooling system is able to reduce the operating temperature of the cells by 16.5°C on average, and it has a better cooling effect when the temperature of the cells becomes higher. In addition, daily electrical yields of the PV module will grow 80Wh, achieving an increment of 8.3%. However, variable environmental conditioning has impacts on gas chamber expansion, so does on water pumping and the cooling effect. Therefore, it is meaningful to evaluate annual performance of the solar-driven water cooling system under a stable environmental condition.

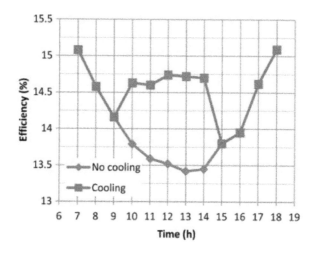

FIGURE 10: Comparison of the efficiencies between with cooling and without cooling to the PV panel.

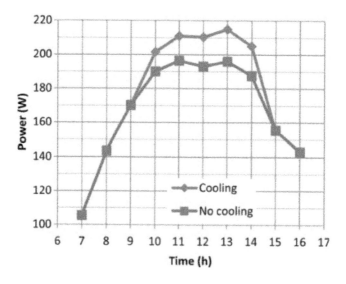

FIGURE 11: Comparison of the power outputs between with cooling and without cooling to PV panel.

6.4.3 MONTHLY WATER CONSUMPTION BY THE SYSTEM

The supply of the rainwater depends on the gas expansion in a chamber, which varies with solar radiation and ambient air temperature. As shown in Figure 12, in January and December, little water can be pumped by this device; however, in June, intensive solar radiations and high air temperatures make the device to pump 110.8 l of water to PV panel for cooling in each day. According to solar radiation and rainwater supply, the system was designed to work between April and September.

For a well-constructed roof, the runoff coefficient is usually assumed as 0.8 [21]. Therefore, monthly rainwater collection can be estimated from the following equation:

Rainwater volume = monthly rainfall × catchment area × runoff coefficient

TABLE 3: Comparison between collected rainwater and required rainwater.

Month	Daylight hours	Equivalent sunny day	Operating day ratio	Rainwater collected (l/day)	Required rainwater (l/day)
April	146.32	14.60	0.95	123	89
May	200.9	18.26	1.56	110	104
June	205.1	18.65	1.64	239	111
July	174.96	14.58	0.95	177	107
August	164.24	14.93	1	158	103
September	141.42	14.14	0.89	104	81

It is not efficient and cost-effective to design this solar-driven rainwater cooling device to work every day, especially for the rainy and cloudy days. Thus, equivalent sunny days in each month can be predicted based on an assumption of 10–12 sun hours in a sunny day in different months. In an ideal scenario, sunny days and rainy days occur intermittently and an operating day ratio (number of sunny days/number of rainy days) is calculated to evaluate the relationship between collected rainwater and required rainwater. Table 3 shows that, except in May, the amount of collected rainwater can meet the requirement of the cooling system in each month. A 1000-l water tank allows it to meet the water consumption up to 10 days under the worst-case scenario like continuous sunny days.

6.4.4 ANNUAL ENERGY SAVING AND PAYBACK PERIOD

The certain amount of rain water can cool more PV modules if heat is removed by evaporation, under the premise that rainwater uniformly covers the PV modules. Based on available rainwater in each month, it is estimated that this solar-driven rainwater cooling system can increase 33.4 kWh of electrical yields for a domestic house when six PV modules are applied. To comprehensively analyse the benefits of a new system, the economic analysis of the solar-driven water cooling system is conducted by the extra cost of equipment required to construct this cooling system, against additional energy benefits obtained from the modified PV panels. The total cost of this passive cooling system is estimated as £197., i.e. £80

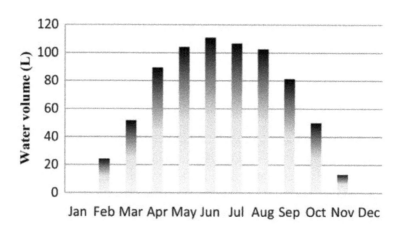

FIGURE 12: Volume of daily pushed rainwater by a gas chamber in different months.

to the cost of the rainwater harvest system and £117 to the cost of the gas expansion chamber, a secondary water tank, pipes and valves. The saving in electrical yields per year equals to £20 when feed-in tariff equals 0.45£/kWh and electricity rate equals 0.145£/kWh [22].

TABLE 4: Payback period analysis.

Year	Power generated at 1% discount rate (kWh)	Electricity supply rate (£) at 2.8% inflation rate	Feed-in-tariff (£) at 2.5% inflation rate	Annual operating saving (£)	Total saving (£)	Initial capital growth (£) at 3% annual interest rate
1	33.40	0.15	0.45	19.87		197.00
2	32.73	0.15	0.46	19.98		202.91
3	32.40	0.15	0.47	20.28		209.00
4	32.06	0.16	0.48	20.59		215.27
5	31.73	0.16	0.50	20.90		221.73
6	31.40	0.17	0.51	21.21		228.38
7	31.06	0.17	0.52	21.53		235.23
8	30.73	0.18	0.53	21.84		242.29
9	30.39	0.18	0.55	22.16		249.55
10	30.06	0.19	0.56	22.48		257.04
11	29.73	0.19	0.58	22.80		264.75
12	29.39	0.20	0.59	23.13		272.69
13	29.06	0.20	0.61	23.45	280.23	280.87
14	28.72	0.21	0.62	23.78	304.01	289.30
15	28.39	0.21	0.64	24.11		297.98
16	28.06	0.22	0.65	24.44		306.92
17	27.72	0.23	0.67	24.77		316.13
18	27.39	0.23	0.68	25.10		325.61
19	27.05	0.24	0.70	25.44		335.38
20	26.72	0.25	0.72	25.77		345.44
21	26.39	0.25	0.74	26.10		355.80
22	26.05	0.26	0.76	26.44		366.48
23	25.72	0.27	0.77	26.77		377.47
24	25.38	0.27	0.79	27.10		388.80
25	25.05	0.28	0.81	27.44	587.50	400.46

A simple payback formula was used to calculate the payback period as follows with an inflation rate of 2.8% being taken into account:

Payback period = (initial cost) / (annual operating saving)

The annual saving in the equation is calculated from:

Annual operating saving = kWh × (electricity rate + feed in tariff)

Assume all the costs of this solar-driven rainwater system are paid up front; the power output of PV discount rate at 1% a year; the electricity inflation rate at 2.8%; the feed-in-tariff inflation rate at 2.5% and annual saving rate at 3%. Based on the assumptions mentioned earlier, the calculation results are shown in Table 4. It can be seen from Table 4 that under this conservative assumption, the payback period is 14 years. Considering that the cost (including water tanks, gas chambers and other equipment) could be reduced with mass production and the additional rainwater collection can be recycled for domestic use in non-operating period, the economic analysis results make this cooling approach quite attractive.

6.5 CONCLUSION

This paper reports a passive cooling system, which can be used for cooling the PV modules on the roof of a domestic house in order to increase electrical efficiency. The simulation results for this cooling system show:

- The influences of the absorbing surface area on the water supply volume are not obvious, whereas a gas chamber with larger volume significantly increases the water supply. However, the actual chamber size should be comprehensively considered with roof area and available rainwater capacity.
- On the design day, the solar-driven rainwater cooling system is able to pump 152 l of water to PV modules. The maximum reduction in the temperature of the cells reaches 19°C and average electrical yield is increased by 8.3%.

- For the solar-driven rainwater cooling system operating between April and September, this cooling system can increase the electricity generation by 33.4 kWh annually.
- The simple payback period of the solar-driven rainwater cooling system was found to be equal to 14 years under a conservative assumption. It still has potential and the initial cost will be reduced if it incorporates with the guttering system.

The most significant point of this approach is that it utilizes rainwater and solar energy to cool the PV panels—improving PV system efficiency with no requirement for additional energy input. The authors believe that it has the potential for further exploration.

REFERENCES

1. Skoplaki E, Palyvos JA. On the temperature dependent of photovoltaic module electrical performance: a review of effective/power correlations. Sol Energy 2009;83:614–24.
2. Notton G, Cristofari C, Mattei M, et al. Modeling of a double-glass photovoltaic module using finite differences. Appl Thermal Eng 2005;25:2854–77.
3. Edenburn MW. Active and passive cooling for concentrating photovoltaic arrays. Conference record, 14th IEEE PVSC, 1980, Washington State Convention Center, USA, pp. 776–776.
4. Araki K, Uozumi H, Yamaguchi M. A simple passive cooling structure and its heat analysis for 500* concentrator PV module. Conference record, 29th IEEE PVSC, 2002,Washington State Convention Center, USA, pp. 1568–71.
5. Tonui JK, Tripanagnostopoulos Y. Improved PV/T solar collectors with heat extraction by forced or natural air circulation. Renew Energy 2007;32: 623–37.
6. Kalogirou AS. Use of TRNSYS for modelling and simulation of a hybrid pv–thermal solar system for Cyprus. Renew Energy 2001;23:247–60.
7. Tripanagnostopoulos Y, Nousia T, Souliotis M, et al. Hybrid photocoltaic/ thermal solar systems. Sol Energy 2002;72:217–34.
8. Krauter S. Increased electrical yield via water flow over the front of photovoltaic panels. Sol Energy Mat Sol Cells 2004;82:131–7.
9. Abdolzadeh M, Ameri M. Improving the effectiveness of a photocoltaic water pumping system by spraying water over the front of photovoltaic cells. Renew Energy 2009;34:91–6.
10. Kordzadeh A. The effects of nominal power of array and system head on the operation of photovoltaic water pimping set with array surface covered by a film of water. Renew Energy 2010;35:1098–102.

11. Furushima K, Nawata Y. Performance evaluation of photovoltaic power generation system equipped with a cooling device utilizing siphonage. Sol Energy Eng 2006;128:146–51.

12. Wilson E. Theoretical and operational thermal performance of a 'wet' crystalline silicon PV module under Jamaican condition. Renew Energy 2009;34:1655–60.

13. Akbarzadeh A, Wadowski T. Heat pipe-based cooling systems for photovoltaic cells under concentrated solar radiation. Appl Therm Eng 1996;16: 81–7.

14. Huang MJ, Eames PC, Norton B. Thermal regulation of building-integrated photovoltaic using phase change materials. Int J Heat Mass Transf 2004;47:2715–33.

15. Biwole P, Eclache P, Kuznik D. Improving the performance of solar panels by the use of phase change materials, World Renewable Energy Congress 2011-Sweden. Photovoltaic Technol 2011;11:2953–60.

16. Gang P, Jie J, Wei H, et al. Performance of photovoltaic solar assisted heat pump system in typical climate zone. J Energy Environ 2007;6:1–9.

17. Xu GY, Deng SM, Zhang XS, et al. Simulation of a photovoltaic/thermal heat pump system having a modified collector/evaporator. Sol Energy 2009;83:1967–76.

18. Song B, Inaba H, Horibe A, et al. Heat, mass and momentum transfer of a water film flowing down a tilted plate exposed to solar irradiation. Int J Therm Sci 1998;38:384–97.

19. Smith CC, Lof G, Jones R. Measurement and analysis of evaporation from an inactive outdoor swimming pool. Sol Energy 1994;53–1:3–7.

20. Sutton Bonington. (2013) Weather data. http://www.metoffice.gov.uk/pub/data/weather/uk/climate/stationdata/suttonboningtondata.txt. (Retrieved on 3 April 2013).

21. Gould J. Rainwater catchment systems for domestic supply: design, construction and implementation. Guildford: Biddles LtdGuildford, 1999, 45–68.

22. FITs. Feed-in tariff scheme. http://www.energysavingtrust.org.uk/ Generating-energy/Getting-money-back/Feed-In-Tariffs-scheme-FITs (2013) (Retrieved on 25 April 2013).

PART IV

SOLAR ENERGY ECONOMICS

CHAPTER 7

A REVIEW OF SOLAR ENERGY: MARKETS, ECONOMICS AND POLICIES

GOVINDA R. TIMILSINA, LADO KURDGELASHVILI, AND PATRICK A. NARBEL

7.1 INTRODUCTION

Solar energy has experienced an impressive technological shift. While early solar technologies consisted of small-scale photovoltaic (PV) cells, recent technologies are represented by solar concentrated power (CSP) and also by large-scale PV systems that feed into electricity grids. The costs of solar energy technologies have dropped substantially over the last 30 years. For example, the cost of high power band solar modules has decreased from about $27,000/kW in 1982 to about $4,000/kW in 2006; the installed cost of a PV system declined from $16,000/kW in 1992 to around $6,000/kW in 2008 (IEA-PVPS, 2007; Solarbuzz, 2006, Lazard 2009). The rapid expansion of the solar energy market can be attributed to a number of supportive policy instruments, the increased volatility of

This chapter was originally published by the World Bank. World Bank Development Research Group, Environment and Energy Team, A Review of Solar Energy: Markets, Economics and Policies, by Govinda R. Timilsina, Lado Kurdgelashvili, and Patrick A. Narbel, Policy Research Working Paper 5845 (October 2011). http://elibrary.worldbank.org/doi/book/10.1596/1813-9450-5845 (accessed 30 June 2014).

fossil fuel prices and the environmental externalities of fossil fuels, particularly greenhouse gas (GHG) emissions.

Theoretically, solar energy has resource potential that far exceeds the entire global energy demand (Kurokawa et al. 2007; EPIA, 2007). Despite this technical potential and the recent growth of the market, the contribution of solar energy to the global energy supply mix is still negligible (IEA, 2009). This study attempts to address why the role of solar energy in meeting the global energy supply mix continues to be so a small. What are the key barriers that prevented large-scale deployment of solar energy in the national energy systems? What types of policy instruments have been introduced to boost the solar energy markets? Have these policies produced desired results? If not, what type of new policy instruments would be needed?

A number of studies, including Arvizu et al. (2011), have addressed various issues related to solar energy. This study presents a synthesis review of existing literature as well as presents economic analysis to examine competitiveness solar energy with fossil energy counterparts. Our study shows that despite a large drop in capital costs and an increase in fossil fuel prices, solar energy technologies are not yet competitive with conventional technologies for electricity production. The economic competitiveness of these technologies does not improve much even when the environmental externalities of fossil fuels are taken into consideration. Besides the economic disadvantage, solar energy technologies face a number of technological, financial and institutional barriers that further constrain their large-scale deployment. Policy instruments introduced to address these barriers include feed in tariffs (FIT), tax credits, capital subsidies and grants, renewable energy portfolio standards (RPS) with specified standards for solar energy, public investments and other financial incentives. While FIT played an instrumental role in Germany and Spain, a mix of policy portfolios that includes federal tax credits, subsidies and rebates, RPS, net metering and renewable energy certificates (REC) facilitated solar energy market growth in the United States. Although the clean development mechanism (CDM) of the Kyoto Protocol has helped the implementation of some solar energy projects, its role in promoting solar energy is very small as compared to that for other renewable energy technologies because of cost competitiveness. Existing studies we reviewed indicate that the share of solar energy in global energy supply mix could exceed 10% by 2050. This would still be a small share of total energy supply and a small share of renewable supply if the carbon intensity of the global energy system were reduced by something on the order of 75%, as many have argued is necessary to stem the threat of global warming.

The paper is organized as follows. Section 2 presents the current status of solar energy technologies, resource potential and market development. This is followed by economic analysis of solar energy technologies, including sensitivities on capital cost reductions and environmental benefits in Section 3. Section 4 identifies the technical, economic, and institutional barriers to the development and utilization of solar energy technologies, followed by a review of existing fiscal and regulatory policy approaches to increase solar energy development in Sections 5 and 6, including potential impacts of greenhouse gas mitigation policies on the deployment of solar energy technologies. Finally, key conclusions are drawn in Section 7.

7.2 CURRENT STATUS OF SOLAR ENERGY TECHNOLOGIES AND MARKETS

7.2.1 TECHNOLOGIES AND RESOURCES

Solar energy refers to sources of energy that can be directly attributed to the light of the sun or the heat that sunlight generates (Bradford, 2006). Solar energy technologies can be classified along the following continuum: 1) passive and active; 2) thermal and photovoltaic; and 3) concentrating and non-concentrating. Passive solar energy technology merely collects the energy without converting the heat or light into other forms. It includes, for example, maximizing the use of day light or heat through building design (Bradford, 2006; Chiras, 2002).

In contrast, active solar energy technology refers to the harnessing of solar energy to store it or convert it for other applications and can be broadly classified into two groups: (i) photovoltaic (PV) and (ii) solar thermal. The PV technology converts radiant energy contained in light quanta into electrical energy when light falls upon a semiconductor material, causing electron excitation and strongly enhancing conductivity (Sorensen, 2000).

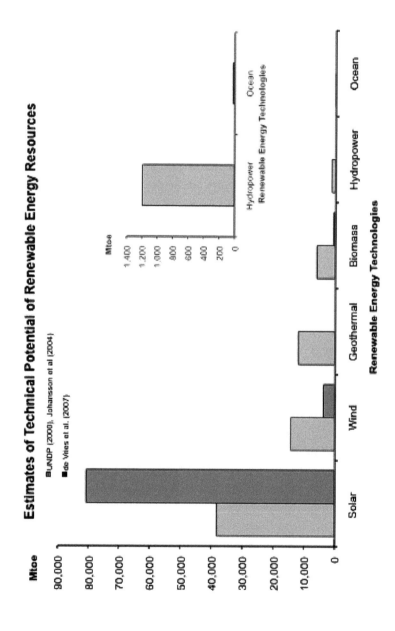

FIGURE 1: Technical potential of renewable energy technologies

Two types of PV technology are currently available in the market: (a) crystalline silicon-based PV cells and (b) thin film technologies made out of a range of different semi-conductor materials, including amorphous silicon, cadmium-telluride and copper indium gallium diselenide. Solar thermal technology uses solar heat, which can be used directly for either thermal or heating application or electricity generation. Accordingly, it can be divided into two categories: (i) solar thermal non-electric and (ii) solar thermal electric. The former includes applications as agricultural drying, solar water heaters, solar air heaters, solar cooling systems and solar cookers (e.g. Weiss et al., 2007); the latter refers to use of solar heat to produce steam for electricity generation, also known as concentrated solar power (CSP). Four types of CSP technologies are currently available in the market: Parabolic Trough, Fresnel Mirror, Power Tower and Solar Dish Collector (Muller-Steinhagen and Trieb, 2004; Taggart 2008a and b; Wolff et al., 2008).

Solar energy technologies have a long history. Between 1860 and the First World War, a range of technologies were developed to generate steam, by capturing the sun's heat, to run engines and irrigation pumps (Smith, 1995). Solar PV cells were invented at Bell Labs in the United States in 1954, and they have been used in space satellites for electricity generation since the late 1950s (Hoogwijk, 2004). The years immediately following the oil-shock in the seventies saw much interest in the development and commercialization of solar energy technologies. However, this incipient solar energy industry of the 1970s and early 80s collapsed due to the sharp decline in oil prices and a lack of sustained policy support (Bradford, 2006). Solar energy markets have regained momentum since early 2000, exhibiting phenomenal growth recently. The total installed capacity of solar based electricity generation capacity has increased to more than 40 GW by the end of 2010 from almost negligible capacity in the early nineties (REN21, 2011).

Solar energy represents our largest source of renewable energy supply. Effective solar irradiance reaching the earth's surface ranges from about $0.06kW/m^2$ at the highest latitudes to $0.25kW/m^2$ at low latitudes. Figure 1 compares the technically feasible potential of different renewable energy options using the present conversion efficiencies of available technologies. Even when evaluated on a regional basis, the technical potential of solar energy in most regions of the world is many times greater than current total primary energy consumption in those regions (de Vries et al. 2007).

Table 1 presents regional distribution of annual solar energy potential along with total primary energy demand and total electricity demand in year 2007. As illustrated in the table, solar energy supply is significantly greater than demand at the regional as well as global level.

TABLE 1: Annual technical potential of solar energy and energy demand (Mtoe)

Region	Minimum technical potential	Maximum technical potential	Primary energy demand (2008)	Electricity demand (2008)
North America	4,322	176,951	2,731	390
Latin America & Caribbean	2,675	80,834	575	74
Western Europe	597	21,826	1,822	266
Central and Eastern Europe	96	3,678	114	14
Former Soviet Union	4,752	206,681	1,038	92
Middle East & North Africa	9,839	264,113	744	70
Sub-Saharan Africa	8,860	227,529	505	27
Pacific Asia	979	23,737	702	76
South Asia	907	31,975	750	61
Centrally Planned Asia	2,746	98,744	2,213	255
Pacific OECD	1,719	54,040	870	140
Total	37,492	1,190,108	12,267	1,446

Note: The minimum and maximum reflect different assumptions regarding annual clear sky irradiance, annual average sky clearance, and available land area. Source: Johansson et al. (2004); IEA (2010)

Kurokawa et al. (2007) estimate that PV cells installed on 4% of the surface area of the world"s deserts would produce enough electricity to meet the world"s current energy consumption. Similarly, EPIA (2007) estimates that just 0.71% of the European land mass, covered with current PV modules, will meet the continent's entire electricity consumption. In many regions of the world 1 km^2 of land is enough to generate more than 125 gigawatt hours (GWh) of electricity per year through CSP technology. In China, for example, 1% (26,300 km^2) of its "wasteland" located in the northern and western regions, where solar radiation is among the highest in the country, can generate electricity equivalent to 1,300 GW—about double the country's total generation capacity projected for year 2020 (Hang et al, 2007). In the United States, an area of 23,418 km^2 in the sunnier southwestern part of the country can match the present generating capacity of 1,067 GW (Mills and Morgan, 2008).

7.2.2. CURRENT MARKET STATUS

The installation of solar energy technologies has grown exponentially at the global level over the last decade. For example, as illustrated in Figure 2(a), global installed capacity PV (both grid and off-grid) increased from 1.4 GW in 2000 to approximately 40 GW in 2010 with an average annual growth rate of around 49% (REN21, 2011). Similarly, the installed capacity of CSP more than doubled over the last decade to reach 1,095MW by the end of 2010. Non-electric solar thermal technology increased almost 5 times from 40 GW$_{th}$ in 2000 to 185 GW$_{th}$ in 2010 (see Figure 3). The impetus behind the recent growth of solar technologies is attributed to sustained policy support in countries such as Germany, Italy United States, Japan and China.

7.2.2.1 SOLAR PV

By December 2010, global installed capacity for PV had reached around 40 GW of which 85% grid connected and remaining 15% off-grid (REN21, 2010). This market is currently dominated by crystalline silicon-based PV cells, which accounted for more than 80% of the market in 2010. The remainder of the market almost entirely consists of thin film technologies that use cells made by directly depositing a photovoltaic layer on a supporting substrate.

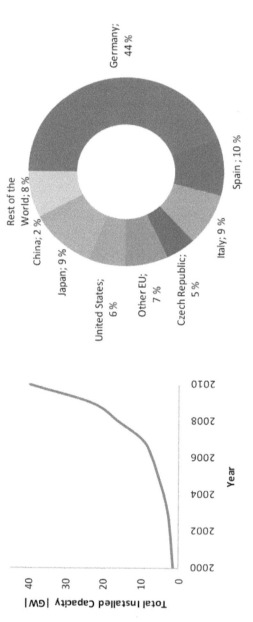

(a) Trend of global installed capacity   (b) Country share in the global installation in 2010

Source: REN21, 2011

FIGURE 2: Total Installed Capacity of PV at the Global Level

As illustrated in Figure 2b, a handful of countries dominate the market for PV. However, a number of countries are experiencing a significant market growth. Notably, Czech Republic had installed nearly 2 GW of solar PV by December 2010 (REN21, 2011), up from almost zero in 2008. India had a cumulative installed PV capacity of 102 MW (EPIA, 2011) and China had a cumulative capacity of 893 MW at the end of 2010.

Two types of PV systems exist in the markets: grid connected or centralized systems and off-grid or decentralized systems. The recent trend is strong growth in centralized PV development with installations that are over 200 kW, operating as centralized power plants. The leading markets for these applications include Germany, Italy, Spain and the United States. After exhibiting poor growth for a number of years, annual installations in the Spanish market have grown from about 4.8 MW in 2000 to approximately 950 MW at the end of 2007 (PVRES 2007) before dropping to 17 MW in 2009 and bouncing back to around 370 MW in 2010 (EPIA, 2011). The off-grid applications (e.g., solar home systems) kicked off an earlier wave of PV commercialization in the 1970s, but in recent years, this market has been overtaken by grid-connected systems. While grid-connected systems dominate in the OECD countries, developing country markets, led by India and China, presently favor off-grid systems. This trend could be a reflection of their large rural populations, with developing countries adopting an approach to solar PV that emphasizes PV to fulfill basic demands for electricity that are unmet by the conventional grid.

7.2.2.2 CONCENTRATED SOLAR POWER (CSP)

The CSP market first emerged in the early 1980s but lost pace in the absence of government support in the United States. However, a recent strong revival of this market is evident with 14.5 GW in various stages of development across 20 countries and 740 MW of added CSP capacity between 2007 and 2010 While many regions of the world, for instance, Southwestern United States, Spain, Algeria, Morocco, South Africa, Israel, India and China, provide suitable conditions for the deployment of CSP, market activity is mainly concentrated in Southwestern United States and Spain, both of which are supported with favorable policies, investment tax credits and feed-in tariffs (Wolff et al. 2008). Currently, several projects around the world are either under construction, in the planning stages, or undergoing feasibility studies and the market is expected to keep growing at a significant pace (REN21, 2011).

7.2.2.3 SOLAR THERMAL FOR HEATING AND COOLING

The total area of installed solar collectors (i.e., non-electric solar thermal) amounted to 185 GWth by early 2010 (REN21, 2011). Of which China, Germany, Turkey and India accounted for 80.3%, 3.1%, 1.8% and 1.1% respectively. The remaining 13.7% was accounted for other 40 plus countries including the USA, Mexico, India, Brazil, Thailand, South Korea, Israel, Cyprus, Ethiopia, Kenya, South Africa, Tunisia, and Zimbabwe. Three types of solar collectors (i.e., unglazed, glazed flat-plate and evacu-

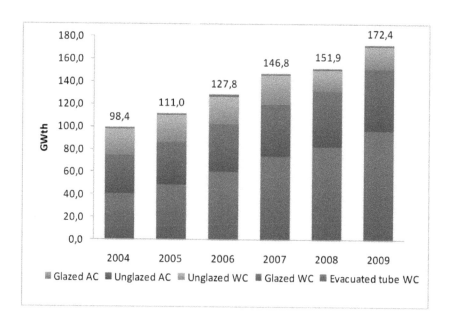

FIGURE 3: Installed Capacity of Solar Thermal Systems. Source: Weiss et al. (2005 to 2011 Issues). WC is water collector and AC is air collector.

ated tube) are found in the market. By the end of 2009, of the total installed capacity of 172.4 GW_{th}, 32% was glazed flat-plate collectors; 56% was evacuated tube collectors; 11% was unglazed collectors; and the remaining 1% was glazed and unglazed air collectors (Weiss et al., 2011).The market for solar cooling systems remains small although it is growing fast. An estimated 11 systems were in operation worldwide by the end of 2009 (REN21, 2011). The use of solar thermal non-electric technologies varies greatly in scale as well as type of technology preferred. For instance, the market in China; Taiwan, China; Japan; and Europe is dominated by glazed flat-plate and evacuated tube water collectors. On the other hand, the North American market is dominated by unglazed water collectors employed for applications such as heating swimming pools.

7.3 THE ECONOMICS OF SOLAR ENERGY

There is a wide variety of solar energy technologies and they compete in different energy markets, notably centralized power supply, grid-connected distributed power generation and off-grid or stand-alone applications. For instance, large-scale PV and CSP technologies compete with technologies seeking to serve the centralized grid. On the other hand, small-scale solar energy systems, which are part of distributed energy resource (DER) systems, compete with a number of other technologies (e.g., diesel generation sets, off-grid wind power etc.). The traditional approach for comparing the cost of generating electricity from different technologies relies on the "levelized cost" method. The levelized cost (LCOE) of a power plant is calculated as follows:

$$LCOE = \frac{OC}{CF \times 8760} \times CRF + OMC + FC \quad \text{with } CRF = \frac{r \times (1+r)^T}{(1+r)^T - 1}$$

where OC is the overnight construction cost (or investment without accounting for interest payments during construction); OMC is the series of annualized operation and maintenance (O&M) costs; FC is the series of annualized fuel costs; CRF is the capital recovery factor; CF is the capacity factor; r is the discount rate and T is the economic life of the plant.

In this section, we discuss the economics of grid connected PV and CSP under various scenarios. One of the main challenges to the economic analysis of power generation technologies is the variation in cost data across technology type, size of plant, country and time. Since fuel costs are highly volatile and capital costs of solar technologies are changing every year, an economic analysis carried out in one year might be outdated the next year. Nevertheless, the analysis presented here could help illustrate the cost competitiveness of solar energy technologies with other technologies at present.

We have taken data from various sources including Lazard (2009), NEA/IEA (2005, 2010), EIA (2007, 2009) and CPUC (2009). The data were available for different years, so we adjusted them using the GDP deflator and expressed them in 2008 prices for our analysis. Moreover, the existing calculations of LCOE for a technology vary across studies as they use different economic lives, capacity factors and discount rates. Some studies account for financial costs (e.g., taxes and subsidies) (Lazard, 2009; CPUC, 2009), while others include only economic costs (NEA/IEA, 2005, 2010). Therefore, we have taken the maximum and minimum values of overnight construction costs for each technology considered here from the existing studies to reflect the variations in overnight construction costs, along with the corresponding O&M and fuel costs, and applied a uniform 10% discount rate and 2.5% fuel price and O&M costs escalation rate to cost data from all the studies. Since our focus is on economic analysis, taxes, subsidies or any types of capacity credits are excluded. Please see Table 2 for key data used in the economic analysis.

Figure 4 presents the results of the levelized cost analysis. Although the costs of solar energy have come down considerably and continue to fall, the levelized costs of solar energy are still much higher compared to conventional technologies for electricity generation, with the exception of gas turbine. For example, the minimum values of levelized cost for solar technologies (US$192/MWh for PV and US$194/MWh for CSP) are more than four times as high as the minimum values of the levelized cost of supercritical coal without carbon capture and storage (US$43/MWh). Among renewable energy technologies, wind and hydropower technologies are far more competitive with fossil fuel and nuclear power plants.

TABLE 2: Key Data Used in Economic Analysis

Technology		Overnight Construction Cost (US$/kW)	Plant Economic Life (years)	Capacity Factor (%)	Source
Solar PV	Min	2878	25	21	NEA/IEA
	Max	7381	25	20	NEA/IEA
Solar CSP	Min	4347	25	34	NEA/IEA
	Max	5800	20	26	Lazard
Wind	Min	1223	25	27	NEA/IEA
	Max	3716	25	23	NEA/IEA
Gas CC	Min	538	30	85	NEA/IEA
	Max	2611	30	85	NEA/IEA
Gas CT	Min	483	25	85	NEA/IEA (2005)
	Max	1575	20	10	Lazard
Hydro	Min	757	80	34	NEA/IEA
	Max	3452	20	50	CPUC
IGCC w CSS*	Min	3569	40	85	NEA/IEA
	Max	6268	40	85	NEA/IEA
Supercritical^	Min	1958	40	85	NEA/IEA
	Max	2539	40	85	NEA/IEA
Nuclear	Min	3389	60	20	EIA
	Max	8375	20	90	Lazard

*Note: * IGCC with carbon capture and storage. ^Supercritical coal.*

The difference between the minimum and maximum values for the levelized costs of solar energy technologies (and also other energy technologies) are wide due mainly to large variations in overnight construction costs and to different capacity factors. For example, the overnight construction costs of grid connected solar PV system vary from US$2,878/kW to US$7,381/kW (NEA/IEA, 2010). Similarly, the overnight construction costs of CSP vary from US$4,347/kW (NEA/IEA, 2010) to US$5,800/kW (Lazard, 2009). The capacity utilization factor of simple cycle gas turbine varies from 10% (Lazard, 2009) to 85% (NEA/IEA, 2010). Furthermore, very different economic lives are assumed for hydro, coal and nuclear plants.

FIGURE 4: Levelized Cost of Electricity Generation by Technology (2008US\$/MWh) Note: * IGCC with carbon capture and storage. ^Supercritical coal.

(a)

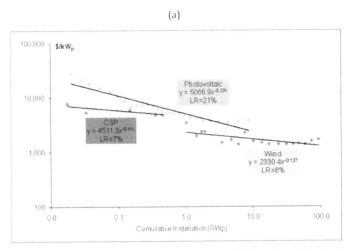

Sources: Earth Policy Institute (2009); DOE (2008b); Stoddard et al. (2007); Charls et al. (2005); Winter (1991)

(b)

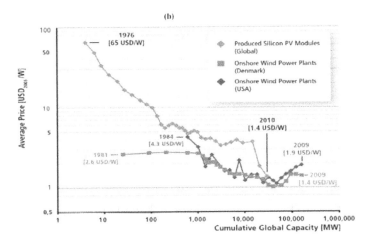

Sources: Arvizu et. al (2011)

FIGURE 5: Experience Curves of Renewable Electric Technologies

It is also interesting to observe the contributions of various cost components (e.g., capital, O&M and fuel costs) to levelized cost. While capital cost accounts for more than 80% of the levelized cost for renewable energy technologies, it accounts for less than 60% in conventional fossil fuel technologies (e.g., coal, gas combined cycle). Fuel costs are the major components in most fossil fuel technologies

Using the concept of experience or learning curves which plot cost as a function of cumulative production on a double-logarithmic scale, implying a constant relationship between percentage changes in cost and cumulative output11, existing studies (e.g., Kannan et al., 2006; Hertlein et al., 1991; EWEA, 2008; Ackerman and Erik, 2005; Dorn, 2007, 2008; Neij, 2008), expect significant reductions in the capital costs of solar energy technologies (see Figure 5a). The cost of solar PV has been declining rapidly in the past, compared not only to conventional technologies such as coal and nuclear, but also to renewable technology such as wind. The 2011 Special Report on Renewable Energy Carried out by Intergovernmental Panel on Climate Change (Arvizu et. al (2011) has also demonstrates reduction in costs of solar and wind power along with their cumulative installed capacity (see Figure 5b). The "learning rate" of solar PV, CSP and wind are 21%, 7%, and 8%, respectively (Nemet, 2007; Beinhocker et al., 2008).

Considering the declining trend of capital costs as discussed above, we analyzed the levelized costs of solar energy technologies when their capital costs drop by 5% to 25% from the present level. Figure 6 shows how the levelized cost of solar thermal trough, solar thermal tower, photovoltaic thin-film and photovoltaic crystalline would decline if their capital cost requirements were to fall by up to 25% and how those costs would compare to the maximum levelized costs of traditional electricity generation plants. As illustrated in the figure, the minimum values of levelized cost of any solar technologies, including tower type CSP, which is currently the least costly solar technology, would be higher than the maximum values of levelized costs of conventional technologies for power generation (e.g., nuclear, coal IGCC, coal supercritical, hydro, gas CC) even if capital costs of solar energy technologies were reduced by 25%.

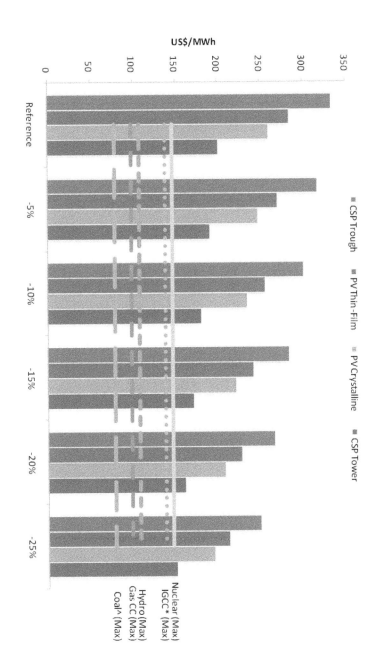

FIGURE 6: Sensitivity of levelized costs of solar technologies to their capital cost reduction

Since fossil fuels such as coal and gas produce negative externalities at the local level (e.g., local air pollution) as well at the global level (e.g., GHG emissions), whereas solar energy technologies do not, it would be unfair to compare solar energy technologies with fossil fuel technologies without accounting for those externalities. Hence, we further analyze the levelized costs of electricity generation technologies, developing a framework to capture some of those external costs. The framework accounts for the environmental damage costs of fossil fuels, particularly climate change damage costs. Damage costs of local air pollution are not included due to a lack of data. Since obtaining actual values of damage costs of emissions from different fossil fuel technologies is highly complex, we employed a sensitivity analysis by considering various values of damage costs ranging from US\$0/tCO$_2$ to US\$100/tCO$_2$. Figure 7 plots the levelized costs of various technologies against the climate change damage costs. The figure demonstrates that the minimum values of levelized costs of solar energy technologies would be higher than the maximum values of the levelized costs of fossil fuel technologies even if the climate change damage costs of 100/tCO$_2$ are imputed to fossil fuel technologies. In other words, even if we assign a climate change damage cost of US\$100/tCO$_2$ to fossil fuel technologies, solar energy technologies would still presently be economically unattractive as compared to fossil fuel technologies.

The analysis above shows that climate change mitigation benefits would not be sufficient to make solar energy technologies economically attractive. However, solar energy technologies also provide additional benefits, which are not normally excluded from traditional economic analysis of projects. For example, as a distributed energy resource available nearby load centers, solar energy could reduce transmission and distribution (T&D) costs and also line losses. Solar technologies like PV carry very short gestation periods of development and, in this respect, can reduce the risk valuation of their investment (Byrne et al., 2005b). They could enhance the reliability of electricity service when T&D congestion occurs at specific locations and during specific times. By optimizing the location of generating systems and their operation, distributed generation resources such as solar can ease constraints on local transmission and distribution systems (Weinberg et al., 1991; Byrne et al., 2005b). They can also protect consumers from power outages. For example, voltage surges of a mere millisecond can cause "brownouts," causing potentially large losses to consumers whose operations require high quality power supply. They carry the potential to significantly reduce market uncertainty accompanying bulk power generation. Because of their modular nature and smaller scale (as opposed to bulk power generation), they could reduce the risk of over shooting demand, longer construction periods, and technological obsolescence (Dunn, 2000 quoted in Byrne et al., 2005b: 14). Moreover, the peak generation time of PV systems often closely matches peak loads for a typical day so that investment in power generation, transmission, and distribution may be delayed or eliminated (Byrne et al., 2005b). However, developing a framework to quantify all these benefits is beyond the scope of this study.

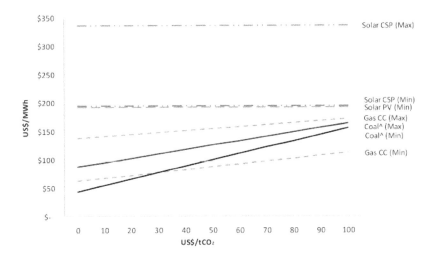

FIGURE 7: Economic attractiveness of solar technologies when environmental damages of fossil fuel technologies are accounted

7.4 ESTIMATED FUTURE GROWTH OF SOLAR ENERGY AND BARRIERS TO REALIZING GROWTH

Advocates of solar energy claim that it will play a crucial role in meeting future energy demand through clean energy resources. Existing projections of long-term growth (e.g., until 2050) of solar energy vary widely based on a large number of assumptions. For example, Arvizu et al. (2011) argue that expansion of solar energy depends on global climate change mitigation scenarios. In the baseline scenario (i.e., in the absence of climate change mitigation policies), the deployment of solar energy in 2050 would vary from 1 to 12 EJ/yr. In the most ambitious scenario for climate change mitigation, where CO_2 concentrations remain below 440 ppm by 2100, the contribution of solar energy to primary energy supply could reach 39 EJ/yr by 2050.

EPIA/Greenpeace (2011) produces the most ambitious projections of future PV installation. The study argues that if existing market supports are continued and additional market support mechanisms are provided, a dramatic growth of solar PV would be possible, which will lead to worldwide PV installed capacity rising from around 40 GW in 2010 to 1,845 GW by 2030. The capacity would reach over 1000 GW in 2030 even with a lower level of political commitment.

A study jointly prepared by Greenpeace International and the European Renewable Energy Council (Teske et al., 2007) projects that installed global PV capacity would expand to 1,330 GW by 2040 and 2,033 GW by 2050. A study by the International Energy Agency (IEA, 2008) estimates solar power development potential under two scenarios that are differentiated on the basis of global CO_2 emission reduction targets. In the first scenario, where global CO_2 emissions in 2050 are restricted at 2005 level, global solar PV capacity is estimated to increase from 11 GW in 2009 to 600 GW by 2050. In the second scenario, where global CO_2 emissions are reduced by 50% from 2005 levels by 2050, installed capacity of solar PV would exceed 1,100 GW in 2050.

Like solar PV, projections are available for CSP technology. A joint study by Greenpeace, the European Solar Thermal Power Industry (ES-TIA) and the International Energy Agency projects that global CSP capacity would expand by one hundred-fold to 37 GW by 2025 and then sky-rocket to 600 GW by 2040 (Greenpeace et al., 2005). Teske et al. (2007) project that global CSP capacity could reach 29 GW, 137 GW and 405 GW in 2020, 2030 and 2050, respectively. IEA (2008) projects that CSP capacity could reach 380 GW to 630 GW, depending on global targets for GHG mitigation. In the case of solar thermal energy, the global market could expand by tenfold to approximately 60 million tons of oil equivalent (Mtoe) by 2030 (IEA World Energy Outlook 2006). A more optimistic scenario from the European Renewable Energy Council (2004) projects that solar thermal will grow to over 60 Mtoe by 2020, and that the market will continue to expand to 244 Mtoe by 2030 and to 480 Mtoe, or approximately 4% of total global energy demand, by 2040. It would be also relevant to envisage the contribution of solar energy to the global energy supply mix. According to EREC (2004), renewable energy is expected to supply nearly 50% of total global energy demand by 2040. Solar energy alone is projected to meet approximately 11% of total final energy consumption, with PV supplying 6%, solar heating and cooling supplying 4% and CSP supplying 1% of the total. Shell (2008) shows that if actions begin to address the challenges posed by energy security and environmental pollution, sources of energy other than fossil fuels account for over 60% of global electricity consumption, of which one third comes from solar energy. In terms of global primary energy mix, solar energy could occupy up to 11% by 2050.

Notwithstanding these optimistic projections, the existing literature identifies a range of barriers that constrains the deployment of solar energy technologies for electricity generation and thermal purposes. These barriers can be classified as technical, economic, and institutional and are presented in Table 3. Technical barriers vary across the type of technology. For example, in the case of PV, the main technical barriers include low conversion efficiencies of PV modules; performance limitations of system components such as batteries and inverters; and inadequate supply of raw materials such as silicon. In the case of stand-alone PV systems, storage is an important concern, as is the shorter battery life compared to that of the module. Furthermore, safe disposal of batteries becomes difficult in the absence of a structured disposal/recycling process. With regard to solar thermal applications, there are two main technical barriers.

TABLE 3: Barriers to the Development and Deployment of Solar Energy Technologies

	PV	Solar Thermal
Technical Barriers	• The efficiency constraint: 4% to 12% (for thin film) and under 22% (for crystalline) in the current market (EPIA/Greenpeace, 2011). • Performance limitations of balance of system (BOS) components such as batteries, inverters and other power conditioning equipments (Rickerson et al., 2007, Beck and Martinot, 2004; O''Rourke et al., 2009). • Silicon supply: strong demand for PV in 2004 and 2005 outpaced the supply and partly stalled the growth of solar sector (Wenzel, 2008; PI, 2006). • Cadmium and tellurium supply for certain thin film cells: these two components are by-products from respectively the zinc mining and copper processing and their availability depends on the evolution of these industries (EPIA/Greenpeace, 2011).	• Heat carrying capacity of heat transfer fluids. • Thermal losses and energy storage system issues with CSPs (Herrmann et al., 2004; IEA, 2006a). • Supply orientation in the design of solar water heaters when product diversity is needed to match diverse consumer demand profiles. • For solar water heating, lack of integration with typical building materials, existing appliances and infrastructure, designs, codes, and standards has hampered widespread application. • In case of central receiver systems the promising technologies such as the molten salt-in-tube receiver technology and the volumetric air receiver technology, both with energy storage system needs more experience to be put for large-scale application (Becker et al., 2000).
Economic Barriers	• High initial capital cost and the related lack of easy and consistent financing options forms one of the biggest barriers primarily in developing countries (Beck and Martinot, 2004). • Investment risks seen as unusually high risks by some financial institutions because of lack of experience with such projects (Goldman et al., 2005; Chaki, 2008 • Cost of BOS is not declining proportional to the decline in module price (Rickerson et al., 2007). • The fragility of solar development partnerships: many PV projects are based on development partnerships and with the early departure of a partner the revenue to complete, operate and maintain the system may falter (Ahiataku-Togobo, 2003).	• High upfront cost coupled with lengthy payback periods and small revenue streams raises creditworthiness risks. • The financial viability of domestic water heating system is low. • Backup heater required in water heating systems to provide reliable heat adds to the cost. • Increasing cost of essential materials like copper make water heating and distribution costly. • Limited rooftop area and lack of building integrated systems limit widespread application.

TABLE 3: *Cont.*

	PV	Solar Thermal
Institutional/ Regulartory Barriers		• The limited capability to train adequate number of technicians to effectively work in a new solar energy infrastructure (Banerjee, 2005; Dayton, 2002). • Limited understanding among key national and local institutions of basic system and finance. • Procedural problems such as the need to work with several public sector agencies (e.g., in India, MNRE, IREDA, the Planning Commisson, and the Ministry of Agriculture and Rural Development) (Radulovic, 2005). • Barriers limiting entry of distributed technology platforms into the grid, including potential for access restrictions by conventional utilities (Margolis and Zuboy, 2006); potential burdens include over-complicated procedures for interconnection, metering and billing (Florida Solar Energy Center, 2000).

They are limits to the heat carrying capacity of the heat transfer fluids and thermal losses from storage systems (Herrmann et al. 2004; IEA 2006a). In addition, as seen in Table 3, there are constraints with regard to system design and integration as well as operating experience for system optimization. For example, lack of integration with typical building materials, designs, codes and standards make widespread application of solar space and water heating applications difficult. In the case of CSP, technologies such as the molten salt-in-tube receiver technology and the volumetric air receiver technology, both with energy storage systems, need more experience to be put forward for large-scale application (Becker et al., 2000). Moreover, solar energy still has to operate and compete on the terms of an energy infrastructure designed around conventional energy technologies.

The economic barriers mainly pertain to initial system costs. Cost comparisons for solar energy technologies by suppliers and users are made against established conventional technologies with accumulated industry experience, economies of scale and uncounted externality costs. Solar energy technologies thus face an "uneven playing field," even as its energy security, social, environmental and health benefits are not internalized in cost calculations (Jacobson & Johnson, 2000). Financing is another critical barrier. Financial institutions consider solar energy technologies to have unusually high risks while assessing their creditworthiness. This is because solar energy projects have a shorter history, lengthy payback periods and small revenue stream (Goldman et al., 2005; Chaki, 2008). This implies higher financial charges (e.g., interest rates) to solar energy projects.

Aside from economic and technical constraints, PV and solar thermal technologies face institutional barriers that reflect considerably the novelty of the technologies. They range from limited capacities for workforce training, to mechanisms for planning and coordinating financial incentives and policies. Inadequate numbers of sufficiently trained people to prepare, install and maintain solar energy systems is another common barrier. In India, for example, the country invested in the training of nuclear physicists and engineers since its independence, while similar requirements for renewable technologies were ignored (Banerjee, 2005).

In some instances, existing laws and regulations could constrain the deployment of solar energy. For example, some applications of small-scale PV systems have had to overcome, cumbersome and inappropriate"interconnection requirements, such as insurance, metering and billing issues, in order to sell excess power generation back into the grid (Florida Solar Energy Center, 2000). However, these potential constraints can become binding only when other policies in place induce or require use of solar energy in order to overcome its higher cost. Even if interconnection were to be simplified, grid based electricity suppliers would still have to address challenges of integrating significant quantities of episodic, non-dispatchable solar power into the grid (or the high cost of current storage options).

7.5 POTENTIAL POLICY INSTRUMENTS TO INCREASE SOLAR ENERGY DEVELOPMENT

As illustrated earlier, by and large solar energy technologies are not yet cost-competitive with conventional energy commodities at either the wholesale or retail levels. Therefore, any significant deployment of solar energy under current technological and energy price conditions will not occur without major policy incentives. A large number of governments have decided to increase solar energy development, using a range of fiscal, regulatory, market and other instruments. In fact, the strong growth in solar energy markets, notably those for grid-connected solar PV and solar thermal water heating, has been driven by the sustained implementation of policy instruments in Europe, the United States and some developing countries to induce or require increased use of solar power.

This section briefly presents key characteristics of policy instruments that support solar energy for both electric and direct heating applications. A large number of policy instruments have been implemented to increase power supplies from solar PV and CSP. The key instruments we highlight here include feed-in-tariffs, investment tax credits, direct subsidies, favorable financing, mandatory access and purchase, renewable energy portfolio standards and public investment. Three rationales are commonly offered for utilizing these policies. One is to encourage the use of low-carbon technology in the absence of a more comprehensive policy for greenhouse gas mitigation, like a carbon tax. The disadvantage of this approach for greenhouse gas mitigation is that it does not create incentives for cost-effective mitigation choices. The second rationale is that expanded investments will ultimately help drive down the costs of those technologies through economies of scale and learning-by-doing. There is clear evidence that scaling-up has driven down unit costs for PV, though not yet to the point that it is cost-effective with conventional alternatives in most cases. CSP is still relatively a pioneer technology with only a few medium-scale investments and no larger-scale investments, though some are planned. It remains to be seen how scale economies and learning-by-doing will lower its costs. The third and most unambiguous rationale is that subsidization of small-scale, off-grid PV (and other renewable energy sources) to bring electricity to remote and poor areas lacking access is a powerful force for stimulating economic development.

7.5.1. FEED-IN-TARIFF

Feed-in-tariff (FiT) refers to a premium payment to new and renewable energy technologies which are relatively expensive or thus not competitive with conventional technologies for electricity generation. The tariff is based on the cost of electricity produced, including a reasonable return on investment for the producer. It thus reduces the risk to potential investors for long-term investments in new and innovative technologies. This policy has been implemented in more than 75 jurisdictions around the world as of early 2010, including in Australia, EU countries, Brazil, Canada, China, Iran, Israel, the Republic of Korea, Singapore, South Africa, Switzerland, the Canadian Province of Ontario and some states in the United States (REN21, 2010). FIT has played a major role in boosting solar energy in countries like Germany and Italy, which are currently leading the world in solar energy market growth. Mendonça and Jacobs (2009) argue that FIT promotes the fastest expansion of renewable electric power at the lowest cost by spreading the costs among all electric utility customers. A study evaluating renewable energy policies in EU countries found that the FIT is the most effective policy instrument to promote solar, wind and biogas technologies (CEC, 2008).

FiTs cover all types of solar energy technologies (e.g., small residential rooftop PV to large scale CSP plants). The tariffs, however, differ across countries or geographical locations, type and size of technology.

For example, German feed-in payments are technology- and scale- specific. It is subdivided by project size, with larger projects receiving a lower feed-in tariff rate in order to account for economies of scale, and by project type, with freestanding systems receiving a low FiT (Sösemann, 2007). The current FITs for solar PV in Germany are 0.43€/kWh for rooftop capacity less than 30 kW; 0.41€/kWh for rooftop capacity between 30 kW and 100 kW; 0.39€/kWh for rooftop capacity between 100 kW and 1MW; 0.33€/kWh for rooftop capacity greater than 1 MW; and 0.32€/kWh for free-standing units (IEA, 2011). Each tariff is eligible for a 20-year fixed-price payment for every kilowatt-hour of electricity generated. Germany's FIT assessment technique is currently based on a "corridor mechanism" (EPIA/Greenpeace, 2011).

This mechanism sets a PV capacity installation growth path which is dependent on the PV capacity installed the year before, and results in a decrease or an increase of the FIT rates according respectively to the percentage that the corridor path was exceeded or unmet. As PV capacity installations were superior than planned by government in 2010, the FIT rates were decreased by 13% on January 1st, 2011 to reflect the decrease in PV costs.

The FiT is regarded as the key driver for recent growth of grid connected solar power, both CSP and grid connected PV. However, some existing studies, such as Couture and Cory (2009), identify several concerns with the FiT. FITs put upward pressure on electricity rates, at least in the near to medium term in order to significantly scale up the deployment of such technologies. FiT policies guaranteeing grid interconnection, regardless of location on the grid, increase transmission costs if projects are sited far from load centers or existing transmission or distribution lines. Similarly, FiT policies designed to periodically adjust to account for changes in technology costs and market prices over time pose a challenge with respect to balancing the purpose of the tariff—increasing utilization of the beneficiary technologies—and fiscal cost, especially as the authorities can only guess at the appropriate tariff adjustments. Changing payment levels increase uncertainties to investors, and political pressures to hold down payments increase overall market risk. In Germany, for example, there was political pressure to cap the policy or speed its rate of decline (Frondel et al., 2008; Podewils, 2007).

7.5.2 INVESTMENT TAX CREDITS

Different types of investment tax credits have been implemented in several jurisdictions around the world to support solar energy. In the United States, for example, the federal government provides an energy investment tax credit for solar energy investments by businesses equal to 30% of expenditures on equipment to generate electricity, to heat or cool and on hybrid solar lighting systems. Besides the investment tax credit, the US federal government provides an accelerated cost-recovery system through depreciation deductions: solar energy technologies are classified as five-year property. In addition, the federal Economic Stimulus Act of 2008, enacted in February 2008, and the American Recovery and Reinvestment Act of 2009, enacted in February 2009, provide a 50% bonus depreciation to solar energy technologies implemented between 2008 and September 2010 and 100% bonus depreciation to solar energy technologies placed in service after September 2010. Residential tax payers may claim a credit of 30% on qualified expenditures on solar energy equipment (e.g., labor costs for onsite preparation, assembly or original system installation). If the federal tax credit exceeds tax liability, the excess amount may be carried forward to the succeeding taxable year until 2016.

The 30% federal tax credits have provided significant leverage to solar energy development in the United Sates, where state governments have further supplemented federal tax incentives with their own programs. For example, the one megawatt CSP project (Sugarno project) installed by Arizona Public Service (APS) in 2006, and the 64 MW Nevada Solar One parabolic trough CSP installed in Boulder City, Nevada in 2007 have largely benefited from the federal tax credit scheme (Canada et al., 2005).

In Bangladesh, the primary driver of the PV market is microcredit finance that led to the substantial growth of privately owned Solar Home Systems (SHS) (IDCOL 2008).

Investment tax credits schemes are criticized for their impacts on government revenues. For example, the investment tax credits in the United States would cost approximately US $907 million over 10 years (Renewable Energy World, July 31, 2008). The tax rebate system in New Jersey would cost $500 million annually to reach the goal; to avoid such high costs, the State Government decided that only systems 10 kW and smaller would qualify for rebates, and systems larger than 10 kW would have to compete in a tradable solar renewable energy credit (SREC) market (Winka, 2006).

7.5.3 SUBSIDIES

Direct subsidies (versus tax credits) are a primary instrument to support solar energy development in most countries. The subsidy could be investment grants or capacity payments, soft loans (e.g., interest subsidies), or output or production based payments. The Spanish government launched a program to provide grants of between €240.40/m² and €310.35/m² in 2000 to solar thermal technologies. In India, capital subsidies initially used, were funded either through donor or government funds. Solar hot water systems, solar cooking systems and concentrating solar cookers receive capital subsidies of, respectively, Rs. 1,500, Rs.1,250 and Rs.2000 per square meter. The primary reliance on capital subsidies was criticized because it incentivized capacity and not necessarily production (Sharma, 2007). In response to these changes, government policy for PV in India has recently been revised. Currently, a production-based subsidy offered by the government has been supplemented by a combined feed-in-tariff of about Rs. 15/kWh for solar PV and solar thermal projects commissioned after March 31st, 2011, for up to 25 years (CERC, 2010). Remote village electrification programs receive even higher levels of subsidies. One such program that aims to establish a single light solar PV system in all non-electrified villages in India by 2012 has 90% of the system cost covered by the government subsidy. In the case of below poverty line (BPL) families, 100% of the system cost will be underwritten by the state governments (MNRE, 2006).

The rebate program for solar PV in California under the California Solar Initiative (CSI) is another example of a subsidy scheme for solar energy. The goal of the $3.3 billion CSI program is to support the development of 3,000 MW of PV in California by 2017 using rebates, also known as Expected Performance-Based Buy-Down (EPBB) based on performance-based incentives (PBI). For systems 50 kW and smaller, the buy-down level is calculated based on expected system performance, taking location and other factors into account. The better the system is projected to perform, the higher the rebate it receives. The level of Buy-Down starts at $2.80 per Watt for the private sector as well as for the public sector and non-profit organizations, which cannot take advantage of the federal tax credit. The rate declines when certain blocks of capacity are reached. Systems over 50 kW are eligible for a five-year PBI which declines in steps similar to the EPBBs. Production incentives of $0.39/kWh for private sector organizations and $0.50/kWh for non-profit and public sector organizations also are offered. Preliminary results indicate that the ambitious target set under the CSI can be reached (CPUC, 2011) with 506 MW already installed by April 2011 and another 403 MW pending. Progress has been most impressive in the residential sector while progresses are slower for the non-residential sector. Previous experience with the program indicated that it would have some trouble achieving its targets without programmatic adjustments (Harris and Moynahan, 2007); however, an increasing rate of new solar installation since 2008 put the program back on track. Although the CSI declines were built into the program to induce efforts to reduce PV costs, it is difficult to match incentive schedules to experience curves (Alsema et al., 2004), and the CSI incentives declined far faster than the 7% annually projected by the program (Go Solar California, 2008). As a result, it remains to be seen whether incentive levels will be too low to sustain market growth in the future, and whether the market will be able to force installation costs low enough to supply attractive systems to customers (Hering, 2008b).

7.5.4 RENEWABLE ENERGY PORTFOLIO (RPS)

Many countries, particularly developed countries, have set penetration targets for renewable energy in total electricity supply mix at the national or state/provincial levels. To meet the targets, electricity suppliers (e.g., utilities, distributors) are required to have certain percentage of their electricity supply coming from renewable energy sources. These standards are commonly known as renewable energy portfolio standards (RPS). The standards can be supplemented with a trading regime where utilities with limited renewable electricity content in their overall supply portfolio, and high cost for renewable energy expansion, can meet their obligation by buying certificates from those with higher renewable electricity content or lower cost of expansion, as illustrated by Tradable Green Certificate (TGC) schemes in Europe. In the United States, 31 out of 50 States have introduced RPS. The standards range from 10% to 40% (Hawaii by 2030). Several states have created an RPS with specific standards for solar energy. The New Jersey RPS required that 6.8% of the electricity sold in the state be renewable by 2008, of which 0.16% was to come from PV. This created a stand-alone market for solar renewable energy credits (SRECs), whose market price was capped through the use of an "alternative compliance payment" (ACP) of $300/MWh. In 2010, New Jersey revised its RPS to require 20.38% of its electricity to come from renewables by 2021. In addition, 2,518 GWh from in-state solar electric facilities must be generated in 2021 and 5,316 GWh in 2026 (DSIRE, 2011). Similarly, Nevada''s RPS mandates that 20% of state electricity come from renewable resource by 2015. Of that, 5% must come from solar power (NREL, 2008). RPS contributed substantially to the realization of large scale CSP plants, such as the 500 MW CSP project in the Imperial Valley in California.

7.5.5 FINANCING FACILITATION

In India, the Shell Foundation worked with two leading banks in India, viz. Canara Bank and Syndicate Bank, to develop renewable energy financing. This initiative helped the banks put in place an interest rate subsidy, marketing support and vendor qualification process. Using the wide network of their branches, the interest subsidies were made available in over 2,000 branch offices in the two states of Kerala and Karnataka. Within two and half years, the programs had financed nearly 16,000 solar home systems, and the subsidies were gradually being phased out. Whereas in 2003 all sales of PV home systems were on a cash and carry basis, by 2006, 50% of sales were financed (Usher and Touhami, 2006).

In Bangladesh, the Rural Electrification and Renewable Energy Development Project established microcredit financed facilities that resulted in the installation of over 970,000 solar-home systems (SHS) between 2003 and May 2011. Having exceeded its expectations, the program now has a target of 1 million SHS systems by 2012 (Uddin and Taplin 2008). This model has been built on the microcredit banking system pioneered by Grameen Bank and now adopted by numerous organizations (IDCOL 2008).

The Spanish government launched a program of low-interest loans for solar thermal applications (7-year loans with interest rates at 2%-3.5% below commercial rates) in 2003 (Institut Català d'Energia, 2003).

7.5.6 PUBLIC INVESTMENT

One of the main drivers of solar energy development in developing countries continues to be direct public investment. Many developing countries host a number of government and/or donor-funded projects to support solar energy under their rural electrification programs. The rapid development of the PV industry and market in China is mainly due to government support, implemented through a number of rural electrification programs. National and local levels programs for rural electrification were the major driving force for solar PV market expansion in China in the late 1990s and early 2000s. The major programs supporting PV programs are Brightness Program Pilot Project, Township Electrification Programs, and China Renewable Energy Development Project. The Brightness Program Pilot Project, launched in 2000, plans to provide electricity to 23 million people in remote areas by 2010, using 2,300 MW of wind, solar PV, wind/PV hybrid and wind/PV/diesel hybrid systems. Inner Mongolia, Gansu and Tibet were selected as pilot provinces, and a RMB 40 million grant was allocated for the project (Ma, 2004). The Township Electrification Programs, launched in 2002, installed 268 small hydro stations and 721 PV, or PV/ wind hybrid systems by 2005 (PMO, 2008). The overall investment was RMB 2.7 billion, and 15.3 MWp of PV systems were installed during the life of the program. The China Renewable Energy Development Project

(REDP), also launched in 2002 and supported by a GEF grant, provided a direct subsidy of US$1.5 per Wp to PV companies to help them market, sell and maintain 10 MWp of PV systems in Qinghai, Gansu, Inner Mongolia, Xinjiang, Tibet and Sichuan.

Developing countries initiated programs with the help of bilateral and multilateral donor agencies are mainly facilitating solar energy development in developing countries. For example, the World Bank has launched a rural power project in the Philippines, aimed at the installation of 135,000 solar systems; totaling 9 MW installed capacity. In addition, the International Finance Corporation finished a 1 MW grid-tied PV with hydro hybrid project in the Philippines (Prometheus Institute, 2007).

In the United States, the federal Energy Policy Act of 2005 established Clean Energy Renewable Bonds (CREBs) as a financing mechanism for public sector renewable energy projects. This legislation originally allocated $800 million of tax credit bonds to be issued between January 1, 2006, and December 31, 2007. The Energy Improvement and Extension Act of 2008 allocated $800 million for new CREBs. The American Recovery and Reinvestment Act of 2009 has allocated an additional $1.6 billion for new CREBs, thereby increasing the size of new CREB allocation to $2.4 billion. In October 2009, the Department of Treasury announced the allocation of $2.2 billion in new CREBs for 805 projects across the country. CREBs may be issued by electric cooperatives, government entities (states, cities, counties, territories, Indian tribal governments or any political subdivision thereof) and by certain lenders. Moreover, the U.S. Department of Agriculture established the Rural Energy for America Program (REAP), which provides grants and loan guarantees for investments in renewable energy systems, energy efficiency improvements and renewable energy feasibility studies. A funding of $255 million has been allocated under this program for the 2009-2012 period.

7.5.7 NET METERING

Net metering is the system where households and commercial establishments are allowed to sell excess electricity they generate from their solar systems to the grid. It has been implemented in Australia, Canada, United

States and some European countries including Denmark, Italy and Spain. In the US, for example, most net metering programs are limited to renewable energy facilities up to 10 kW. In California it could reach up to 1 MW. In Canada, it goes up to 100 kW in Prince Edward Island and 500 kW in Ontario. Most programs only require purchases up to the customer"s total annual consumption, and no payment is offered for any electricity generated above this amount. They receive the retail tariff for their output.

7.5.8 OTHER GOVERNMENT REGULATORY PROVISIONS

In many countries, governments have introduced laws mandating transmission companies and electricity utilities to provide transmission or purchase electricity generated from renewable energy technologies, including solar. In January 2006, China, for example, issued the Renewable Energy Law, mandating utility companies to purchase "in full amounts" renewable energy generated electricity within their domains at a price that includes production cost plus a reasonable profit. The extra cost incurred by the utility will be shared throughout the overall power grid (GOC, 2005). Similarly, in Germany, all renewable energy generators are guaranteed to have priority access to the grid. Electric utilities are mandated to purchase 100% of a grid-connected PV system's output, regardless of whether the system is customer-sited or not.

Government regulations mandating installation of solar thermal systems is the main policy driver for the development of solar thermal applications in many countries (e.g., Spain, Israel). Israel has had a solar water heating obligation for new construction in place since the 1980s, but it did not spread to other countries immediately. In the late 1990s, the City of Berlin proposed to create a similar solar water heating mandate, but was unsuccessful in its attempt. The Spanish city of Barcelona, however, adapted the proposed Berlin mandate, and passed an ordinance in July, 1999, requiring that all new construction or major renovation projects be built with solar water heating (Schaefer, 2006). The original ordinance, which targeted only certain building subsets, such as residential buildings, hotels, and gymnasiums, required that at least 60% of the hot water load be supplied by solar energy. The "Barcelona model" was adopted by 11 other Spanish cities by 2004 (Pujol, 2004), including Madrid, and in 2006, Spain passed a national law requiring solar water heating on new construction and major renovations (ESTIF, 2007).

In China, the Renewable Energy Law requires the government to formulate policies that guide the integration of solar water heaters (SWH) and buildings; real estate developers to provide provisions for solar energy utilization; and residents in existing buildings to install qualified solar energy systems if it does not affect building quality and safety (GOC, 2005). In regions with high solar radiation, hot water intensive public buildings (such as schools and hospitals) and commercial buildings (such as hotels and restaurants) will be gradually mandated for SWH installation. New buildings will need to reserve space for future SWH installation and piping (NDRC, 2008). At provincial and local levels, the governments have issued various policies for SWH promotion; for instance, Jiangsu, Gansu and Shenzhen require buildings of less than 12 floors to be equipped with solar water heaters (Hu, 2006 & 2008).

7.6 IMPLEMENTATION OF POLICIES TO INCREASE SOLAR ENERGY DEVELOPMENT

7.6.1 POLICY MIX

The policy landscape for solar energy is complex with a broad range of policy instruments driving market growth. The rapid market growth of solar energy in Germany and Spain could be attributed to the feed-in-tariff systems that guarantee attractive returns on investment along with the regulatory requirements mandating 100% grid access and power purchase. On the other hand, federal and state incentives, along with regulatory mechanisms such as RPS, get credit for the rapid deployment of solar energy in the United States. In both markets, the policy landscape is in a transitional phase. In Germany, the FiT level is being reduced, whereas in the United States, upfront incentives are being shifted toward performance-based incentives. It is, however, uncertain if the transition will produce expected results. The decrease in the FiT, the primary basis for investors' confidence, could drive investors away from solar energy markets.

The rapid growth of the grid-connected PV and CSP market is largely attributed to a policy suite that guarantees attractive returns on investment, along with regulatory requirements such as grid connectivity and power purchase commitments required to motivate investments. While FITs played an instrumental role in Germany and Italy, a mix of policy portfolios that includes federal tax credits, subsidies and rebates, RPS, net metering and renewable energy certificates (REC) facilitated solar energy market growth in the United States. Similarly, New Jersey developed a policy mix that combined a broad range of federal and state incentives to drive rapid market growth: a policy portfolio consisting of RPS, federal tax credits, grants, drove the rapid growth of the PV market in New Jersey. In the Southwest United States, the combination of excellent solar resources, the 30% federal tax credit, and RPS policies has resulted in a rebirth of solar thermal electric generation. In two of the three states exploring solar thermal electric, the existence of a solar- or distributed generation-specific RPS tier has also played a role in increasing project development.

The capital subsidy was the predominant policy instrument early on in India, but a mix of policy instruments, such as, subsidies, fiscal incentives, preferential tariffs, market mechanisms and legislation, were encouraged later for the deployment of solar energy (MNRE, 2006). For instance, in 2004-05, the subsidy for the solar photovoltaic program varied between 50% and as high as 90% for the "special category states and islands." Similarly, the subsidy for solar photovoltaic water pumping was Rs. 100/Wp and as much as Rs. 135/W in the special category states (Banerjee, 2005). The growing role of private finance has reduced the role of fiscal policy drivers in the overall financing mix for solar power, and capital subsidies have been ratcheted down substantially, except in exceptional cases such as "remote villages and hamlets." India now relies on a mix of mechanisms including various tax and generation-based incentives, renewable purchase obligations, capital subsidies and accelerated depreciation. Yet, the accumulation of incentive programs and the failure to coordinate them is thought to hinder the development of renewable energy resources in India as it results in unnecessary delays and conflicts (ESMAP, 2011a).

In the Philippines, the portfolio of policy instruments includes duty-free importation of equipment, tax credits on domestic capital equipment and services, special realty tax rates, income tax holidays, net operating loss carry-over, accelerated depreciation and exemption from the universal charge and wheeling charges (WWF, 2008).

7.6.2 IMPLEMENTATION CHALLENGES

Sensitivity to policy costs is more significant in developing country markets such as India, China, Brazil, Philippines and Bangladesh than in more developed economies. Thus, a common approach toward renewable energy technologies, seen in developing countries, is to "rationalize development and deployment strategy" (MNRE 2006) of renewable energy technologies. For instance, India planned in its eleventh Five-Year plan (2007-2012) to install 15,000 MW of grid-connected renewable energy and it was widely believed that this market expansion would be driven by wind, micro-hydro and biomass, as the plan recognized that solar PV would be an option only if the prices come down to levels comparable to micro-hydro.

More recently, the National Solar Mission promoting solar power in India has been launched. The first phase (2009-2013) targets increases in the utility grid power from solar sources, including CSP, by over a 1 GW (ESMAP, 2011a). By 2022, 20 GW of solar capacity is to be added in India. The approach to the renewable energy mix in China, Philippines and Bangladesh represents similar priorities of rationalizing the policy costs. In Brazil, as in other developing countries, the minimal policy cost is ensured via technology-specific and reserve energy auctions (ESMAP, 2011b) as the cheapest renewable energy projects are implemented first.

Solar PV is recognized as serving a niche market that is very important in developing countries—electrification of rural and peri-urban areas that do not yet have access to the electri grid. There are vigorous efforts to expand the market for Solar Home Systems (SHS) as a means toward rural electrification. However, rural and peri-urban areas are characterized by low income households that may not be able to afford solar energy technologies unless they are substantially subsidized. Until now, the approach is to provide subsidies either via government funds or through international donors. However, a subsidy is a short-term support, not a long-term solution.

CSP and solar water heating are comparatively cheaper than solar PVs. These could be cost competitive with conventional fuels if existing subsidies to the latter are reduced or removed. However, fossil fuel subsidies are politically sensitive in many countries and their removal might take time. Thus far, CSP has not found much success in a developing country context. Unlike Solar PV, CSP is limited to utility scale applications and as such is often out of consideration in the traditional utility generation market due to current prices. Thus, developing country governments have adopted a cautious policy approach to this market, focusing more on pilot scale projects, as with grid-connected solar PV. Through its National Solar Mission, India is the first developing country to take a step towards the installation of CSP capacity.

Unlike in electric applications, solar heating applications enjoy limited policy support as instruments like FITs and RPS are not applicable for heating applications. Moreover, it is more difficult to measure and verify solar water heating performance, and so performance-based incentives are harder to enact.

7.6.3 SOLAR ENERGY DEVELOPMENT UNDER POLICIES FOR CLIMATE CHANGE MITIGATION

Greenhouse gas mitigation policies and activities help support renewable energy development, including solar energy. Various incentives and mandates designed to trigger GHG mitigation have helped promote solar energy in industrialized countries. In the case of developing countries, the Clean Development Mechanism (CDM) under the Kyoto Protocol has been the main vehicle to promote solar energy under the climate change regime. The CDM allows industrialized countries to purchase GHG reductions achieved from projects in developing countries, where reducing GHG emissions is normally cheaper than in industrialized countries.

As of July 2011, there are 6,416 projects already registered or in the process of registration under the CDM. Of these, 109 projects are solar energy projects with annual emission reduction of 3,570,000 tons of CO_2. Out of these 109 projects, 89 are located in China, South Korea and India.

However, the solar energy projects account for a very small fraction (< 1%) of total emission reductions from the total CDM projects already registered or placed in registration process (UNEP Risoe, 2011).

One reason for the small share of solar energy projects in the global CDM market is cost. As noted, solar energy technologies remain costly, and at present they are not economically competitive with other CDM candidates such as wind power, small hydro, landfill gas, and biomass cogeneration. The high upfront capital investment cannot be recovered even if the revenue generated from sales of emission mitigation at standard (non-subsidized) rates is included along with revenue from electricity sales. In addition, solar energy projects to date come in smaller sizes than other CDM options; transaction costs incurred in various steps during the CDM process (e.g., validation and registration of projects and monitoring, verification and certification of emission reductions) do not vary that much with project size and are often prohibitive for solar energy projects that are already less attractive compared to their competitors.

To increase the share of solar energy projects in the CDM, one approach is to give solar energy technologies some additional premium for other economic and social benefits. However, other technologies can provide these benefits with lower impacts on electricity costs, so the strength of this argument is open to question. The transaction costs of diffused, small-scale solar CDM projects could be reduced by bundling them into single larger projects, as with "programmatic CDM" schemes. Further simplification of CDM registration process for solar energy projects could be accomplished by avoiding additionality screening, as they meet the additionality criterion by default given their costs. With or without CDM, further capacity building in developing countries to enhance technical and managerial skills for market participants is necessary (BMU, 2007).

7.7 CONCLUSIONS

Physically, solar energy constitutes the most abundant renewable energy resource available and, in most regions of the world, its theoretical potential is far in excess of the current total primary energy supply in those regions. Solar energy technologies could help address energy access to rural and remote communities, help improve long-term energy security and help greenhouse gas mitigation.

The market for technologies to harness solar energy has seen dramatic expansion over the past decade—in particular the expansion of the market for grid-connected distributed PV systems and solar hot water systems have been remarkable. Notably, centralized utility scale PV applications have grown strongly in the recent years; off-grid applications are now dominant only in developing markets. Moreover, the market for larger solar thermal technologies that first emerged in the early 1980s is now gathering momentum with a number of new installations as well as projects in the planning stages.

While the costs of solar energy technologies have exhibited rapid declines in the recent past and the potential for significant declines in the near future, the minimum values of levelized cost of any solar technologies, including tower type CSP, which is currently the least costly solar technology, would be higher than the maximum values of levelized costs of conventional technologies for power generation (e.g., nuclear, coal IGCC, coal supercritical, hydro, gas CC) even if capital costs of solar energy technologies were reduced by 25%. Currently, this is the primary barrier to the large-scale deployment of solar energy technologies. Moreover, the scaling-up of solar energy technologies is also constrained by financial, technical and institutional barriers.

Various fiscal and regulatory instruments have been used to increase output of solar energy. These instruments include tax incentives, preferential interest rates, direct incentives, loan programs, construction mandates, renewable portfolio standards, voluntary green power programs, net metering, interconnection standards and demonstration projects. However, the level of incentives provided through these instruments has not been enough to substantially increase the penetration of solar energy in the global energy supply mix. Moreover, these policy instruments can create market inefficiencies in addition to the direct costs of requiring more-costly electricity supplies to be used. While not discussed in this paper, these indirect impacts need to be considered in assessing the full opportunity cost of policies to expand solar power production.

Carbon finance mechanisms, in particular the CDM, could potentially support expansion of the solar energy market. While some changes in the operation of the CDM could increase solar investment, the price of carbon credits required to make solar energy technologies economically competitive with other technologies to reduce GHG emissions would be high.

The fundamental barrier to increasing market-driven utilization of solar technologies continues to be their cost. The current growth of solar energy is mainly driven by policy supports. Continuation and expansion of costly existing supports would be necessary for several decades to enhance the further deployment of solar energy in both developed and developing countries, given current technologies and projections of their further improvements over the near to medium term. Overcoming current technical and economic barriers will require substantial further outlays to finance applied research and development, and to cover anticipated costs of initial investments in commercial-scale improved-technology production capacity.

REFERENCES

1. Ackerman, T., and Morthorst, P. E. (2005). Economic Aspect of Wind Power in Power System. In T. Ackerman (Ed.), Wind Power in Power Systems. The Atrium, West Sussex, England: John Wiley and Sons, Ltd.

2. Ahiataku-Togobo, W. (2003). Challenges of Solar PV for Remote Electrification in Ghana. Accra, Ghana: Renewable Energy Unit, Ministry of Energy, 2003. Retrieved August 20, 2008, from www.zef.de/fileadmin/webfiles/renewables/praesentations/ Ahiataku-Togobo_solar%20PV%20Ghana.pdf

3. Akshay Urja (2008). Special Issue on Solar Energy for Urban and Industrial Applications. Vol. 1. No. 5. March-April 2008. Ministry of New and Renewable Energy, Government of India.

4. Arvizu, D., P. Balaya, L. Cabeza, T. Hollands, A. Jäger-Waldau, M. Kondo, C. Konseibo, V.

5. Meleshko, W. Stein, Y. Tamaura, H. Xu, R. Zilles, 2011: Direct Solar Energy. In IPCC Special Report on Renewable Energy Sources and Climate Change Mitigation [O. Edenhofer, R. Pichs-Madruga, Y. Sokona, K. Seyboth, P. Matschoss, S. Kadner, T. Zwickel, P. Eickemeier, G.Hansen, S. Schlömer, C. v. Stechow (eds)], Cambridge University Press, Cambridge, United Kingdom and New York, NY, USA.

6. Bagnall, D. M., & Boreland, M. (2008). Photovoltaic technologies. Energy Policy, 36(12), 4390-4396.

7. Banerjee, R. (2005). Renewable Energy: Background Paper Submitted to the Integrated Energy Policy Committee, Government of India. Retrieved September 22, 2008, from http://www.whrc.org/policy/COP/India/REMay05%20(sent%20by%20Rangan%20Banerjee).pdf

8. Barnett, A., Kirkpatrick, D., Honsberg, C., Moore, D., Wanlass, M., Emery, K., Schwartz, R., Carlson, D., Bowden, S., Aiken, D., Gray, A., Kurtz, S., Kazmerski, L., Moriarty, T., Steiner, M., Gray, J., Davenport, T., Buelow, R., Takacs, L., Shatz, N., Bortz, J., Jani, O., Goossen, K., Kiamilev, F., Doolittle, A., Ferguson, I., Unger, B., Schmidt, G., Christensen, E., and Salzman, D. (2007). Milestones Toward 50% Efficient Solar Modules. Presented at the 22nd European Photovoltaic Solar Energy Conference, Milan, Italy, 3rd September.

9. Beck, F. and Martinot, E. (2004). Renewable Energy Policies and Barriers. In Cutler Cleveland (Ed), Encyclopedia of Energy, 365-383, San Diego: Academic Press/Elsevier Science.

10. Becker, M., Meinecke, W., Geyer, M. Trieb, F., Blanco, M., Romero, M., and Ferriere, A. (2000). Solar Thermal Power Plants. Prepared for EUREC-Agency. Retrieved September 23, 2008, from www.solarpaces.org/Library/docs/EUREC-Position_Paper_STPP.pdf

11. Beinhocker, E., Oppenheim, J., Irons, B., Lahti, M., Farrell, D., Nyquist, S., Remes, J., Naucler, T., & Enkvist, P. (2008). The carbon productivity challenge: Curbing climate change and sustaining economic growth. Sydney: McKinsey Global Institute, McKinsey & Company.

12. Bradford, T. (2006). Solar Revolution. The Economic Transformation of the Global Energy Industry. Cambridge, MA: The MIT Press.

13. _____. (2008). World PV Market Update and Photovoltaic Markets, Technology, Performance, and Cost to 2015. Proceedings of the Solar Market Outlook: A Day of Data, New York, NY.

14. Byrne, J., Waegel, A., Haney, B., Tobin, D., Alleng, G., Karki, J., and Suarez, J. (2006). Pathways to a U.S. Hydrogen Economy: Vehicle Fleet Diffusion Scenarios. Newark, DE: Center for Energy and Environmental Policy.

15. California Public Utilities Commission (CPUC) (2008). Proposed Decision Filed by ALJ/SIMON/CPUC on 10/29/2008. Retrieved October 29, 2008, from http://docs.cpuc.ca.gov/PD/92913.htm

16. California Public Utilities Commission (CPUC) (2011). California Solar Initiative Annual Program Assessment, June 2011.

17. Central Electricity Regulatory Commission (CERC) (2010). Petition No. 256/2010 (suo.motu) on 09/11/2010. New Dehli, India.

18. Charls, R. P., Davis, K. W., and Smith, J. L. (2005). Assessment of Concentrating Solar Power technology Cost and Performance Forecasts. http://www.trec-uk.org.uk/reports/sargent_lundy_2005.pdf

19. Chaki, C. R. (2008). Use of Solar Energy: Bangladesh Context, Experience of Grameen Shakti. Retrieved November 10, 2008, from www.pksf-bd.org/seminar_fair08/Seminar_day2/GS%20Chitta%20Ranjan%20Chaki.pdf

20. Clean Development Board Executive Board (CDMEB, 2007), Report of the Thirty Second Meeting of the Executive Board of the Clean Development Mechanism, retrieved on January 02 2010 from http://cdm.unfccc.int/EB/032/eb32rep.pdf

21. Commission of the European Communities (CEC, 2008), The support of electricity from renewable energy sources, Commission Staff Working Document, Brussels.
22. Couture, T. and K. Cory (2009), State Clean Energy Policies Analysis (SCEPA) Project: An
23. Analysis of Renewable Energy Feed-in Tariffs in the United States, National Renewable Energy Laboratory, Technical Report NREL/TP-6A2-45551, Colorado.
24. Daymond, C. (2002). PV Focus Group Report. Portland, OR: Energy Trust of Oregon.
25. de Vries B. J. M., van Vuuren, D. P., and Hoogwijk, M. M. (2007). Renewable energy sources: Their global potential for the first-half of the 21st century at a global level: An integrated approach. Energy Policy, 35, 2590-2610.
26. del Río, P., and Gual, M. A. (2007). An Integrated Assessment of the Feed-in Tariff System in Spain. Energy Policy, 35(2), 994-1012.
27. Department of Energy (DOE). (2008). Solar Energy Industry Forecast: Perspectives on U.S. Solar Market Trajectory. Retrieved November 7, 2008, from www1.eere. energy.gov/solar/solar_america/pdfs/solar_market_evolution.pdf
28. Department of Energy (DOE). (2008a). Multi Year Program Plan 2008-2012. Solar Energy Technologies Program, Energy Efficiency and Renewable Energy, US Department of Energy. Retrieved March 26, 2009, from http://www1.eere.energy.gov/solar/pdfs/solar_program_mypp_2008-2012.pdf
29. Department of Energy (DOE). (2008b). Annual Report on U.S. Wind Power Installation, Cost, and Performance Trends: 2007. DOE Energy Efficiency and Renewable Energy. Figure 22, page 21
30. Dienst, C., Fischedick, M., del Valle, S., Bunse, M., Wallbaum, H., and Höllermann, B. (2007). Solar cooling: Using the sun for climiatisation. Wuppertal, Germany: Wuppertal Institute for Climate, Environment and Energy, WISIONS, Promotion of Resource Efficiency Projects (PREP).
31. Dorn, J. G. (2007). Solar Cell Production Jumps 50 Percent in 2007. Retrieved September 12, 2008, from http://www.earth-policy.org/Indicators/Solar/2007.htm
32. _____. (2008). Solar Thermal Power Coming to a Boil. Retrieved September 5, 2008, from http://www.earth-policy.org/Updates/2008/Update73.htm
33. DSIRE. (2011). Modified Accelerated Cost-Recovery System (MACRS) + Bonus Depreciation (2008-2012). Retrieved July 17, 2011, from http://www.dsireusa.org/incentives/?State=US&ee=1&re=1Earth Policy Institute. 2009. Climate, Energy, and Transportation. Retrieved January 10, 2010, from http://www.earth-policy.org/index.php?/data_center/C23/
34. EESI. 2009. Fact Sheet: Concentrated Solar Power. Environmental and Energy Study Institute. August 2009. Washington, D.C.
35. EIA (DOE). (2005). International Electricity Installed Capacity. Retrieved August 10, 2008, from http://www.eia.doe.gov/emeu/international/electricitycapacity.html
36. EIA (DOE). (2008a). Steam Coal Prices for Electricity Generation. Retrieved September 10, 2008, from http://www.eia.doe.gov/emeu/international/stmforelec.html
37. EIA (DOE). (2008b). U.S. Natural Gas Electric Power Price. Retrieved September 8, 2008, from http://tonto.eia.doe.gov/dnav/ng/hist/n3045us3a.htm
38. EIA (DOE). (2008c). U.S Residual Fuel Oil Retail Sales by All Sellers. Retrieved September 8, 2008, from http://tonto.eia.doe.gov/dnav/pet/hist/d300600002A.htm

39. EIA (DOE). (2008d). Uranium Purchased by Owners and Operators of U.S. Civilian Nuclear Power Reactors. Retrieved September 8, 2008, from http://www.eia.doe.gov/cneaf/nuclear/umar/summarytable1.html

40. Emerging Energy Research (2009). Global Concentrated Solar Power Markets and Strategies, 2009-20020. Cambridge, MA: Emerging Energy Research. Retrieved April 20, 2009, from, http://www.emerging-energy.com/user/GlobalConcentratedSolarPowerMarketsandStrategies200920201561467216_pub/SolarCSPTableof-Contents.pdf

41. EPIA/Greenpeace (2008). Solar Generation V – 2008. Greenpeace and European Photovoltaic Industry Association.

42. EPIA/Greenpeace (2011). Solar Generation VI-2011. Greenpeace and European Photovoltaic Industry Association.

43. ESMAP/WB (2008). Study of Equipment Prices in the Energy Sector (Draft). Washington DC: World Bank.

44. ESMAP/WB (2011a). Unleashing the Potential of Renewable Energy in India. Washington DC: World Bank.

45. ESMAP/WB (2011b). Design and Performance of Policy Instruments to Promote the Development of Renewable Energy: Emerging Experience in Selected Developing Countries. Washington DC: World Bank.

46. ESTIF (2007). Solar Thermal Action Plan for Europe: Heating and Cooling from the Sun. Bruxelles, Belgium: European Solar Thermal Industry Federation. Retrieved June 4, 2008, from http://www.estif.org/282.0.html, accessed on June 4, 2008

47. European Commission(EC) (2004). European Research on Concentrated Solar Thermal Energy. EUR 20898, Belgium: Directorate-General for Research Sustainable Energy Systems. Retrieved August 18, 2008, from http://ec.europa.eu/research/energy/pdf/cst_en.pdf See http://europa.eu.in for more related information.

48. European Photovoltaic Industry Association (EPIA). (2004). EPIA roadmap. Brussels, Belgium: European Photovoltaic Industry Association.

49. _____. (2007). PV to Become a Leading World Energy Market. Retrieved November 1, 2008 from http://www.rtcc.org/2007/html/dev_solar_epia.html.

50. European Renewable Energy Council. (2004). Renewable energy scenario to 2040. Brussels, Belgium: European Renewable Energy Council. Retrieved November 25, 2008 from http://www.erec.org/documents/publications/2040-scenario.html.

51. European Wind Energy Association. (EWEA). (1999). Wind Force 10: A Blueprint to Achieve 10% of the World"s Electricity from Wind Power by 2020. London: Forum for Energy and Development and Greenpeace International.

52. EWEA (2008). Pure Power: Wind Energy Scenario up to 2030. European Wind Energy Association, Brussels.

53. Federal Ministry for the Environment, Nature Conservation and Nuclear Safety (BMU). (2007). Renewable Energy and the Clean Development Mechanism. BMU Brochure. Berlin: BMU. Retrieved September 20, 2008, from http://www.bmu.de/english/renewable_energy/downloads/doc/40586.php.

54. Figueres, C. (2005). Policies and Programs under the CDM. Short Note on the COP/MOP1 decision on Programmatic CDM. Retrieved September 22, 2008, from http://www.figueresonline.com/programmaticcdm.htm.

55. First Solar (2009). First Solar Passes $1 per Watt Industry Milestone. Retrieved April 20, 2009, from http://investor.firstsolar.com/phoenix.zhtml?c=201491&p=irol-newsArticle&ID=1259614

56. Florida Solar Energy Center (2000). Florida Photovoltaic Buildings Program: Status Report, Observations, and Lessons Learned. Cocoa, FL: Florida Solar Energy Center.

57. Frondel, M., Ritter, N., and Schmidt, C. M. (2008). Germany's Solar Cell Promotion: Dark Clouds on the Horizon (Ruhr Economic Paper #40). Essen, Germany: Rheinisch-Westfälisches Institut für Wirtschaftsforschung.

58. German Advisory Council on Global Change (WBGU). (2004). World in Transition: Towards Sustainable Energy Systems. Retrieved November 27, 2008 from http://www.wbgu.de/wbgu_publications_annual.html.

59. Geyer, M. (2008, March 4-7). Introducing Concentrated Solar Power on the International Markets: Worldwide Incentives, Policies and Benefits. Proceedings of the 14th Biennial Solar Power and Chemical Energy Systems (SolarPACES) Symposium, Las Vegas, NV.

60. Goett, A., Farmer, R., Moore, D., & Hitchner, R. (2003). Prospects for Distributed Electricity Generation. A Congressional Budget Office Study. Retrieved July 28, 2008, from http://www.cbo.gov/ftpdocs/45xx/doc4552/09-16-Electricity.pdf

61. Goldman, D.P., McKenna J.J., Murphy, L.M. (2005). Financing Projects That Use Clean-Energy Technologies: An Overview of Barriers and Opportunities. NREL/TP-600-38723. Golden, CO: National Renewable Energy Laboratory.

62. Greenpeace, European Solar Thermal Power Industry Association, and SolarPACES. (2005). Concentrated solar thermal power - Now! Retrieved November 24, 2008 from http://www.greenpeace.org/raw/content/international/press/reports/Concentrated-Solar-Thermal-Power.pdf

63. GRI (1999). The Role of Distributed Generation in Competitive Market. GRI Report. Retrived September 19, 2008, from http://www.cbo.gov/doc.cfm?index=4552

64. Held, A., Ragwitz, M., Huber, C., Resch, G., Faber, T., and Vertin, K. (2007). Feed-in Systems in Germany, Spain and Slovenia: A Comparison. Karlsruhe, Germany: Fraunhofer Institut für Systemtechnik und Innovationsforschung.

65. Herrmann, U., Kelly, B., and Price, H. (2004). Two-tank Molten Salt Storage for Parabolic Thorough Solar Power Plants. Energy, 29(5-6), 883-893.

66. Hertlein, H.P., Klaiss, H., and Nitsch, J. (1991). Cost Analysis of Solar power Plants. In C. J. Winter, R. L. Sizmann and L. L. Vant-Hull (Eds.), Solar Power Plants: Fundamentals, Technology, Systems Economics. Berlin, New York, London: Springer-Verlag.

67. Honsberg, C. and Barnett, A. (2008). Achieving a Solar Cell of Greater than 50 Percent: Physics, Technology, Implementation and Milestones. Retrieved September 25, 2008, from http://www.iee.ucsb.edu/event/solar-cell/

68. IDCOL (Infrastructure Development Company Limited) (2008). IDCOL''s Solar Energy Programme. Bangladesh. Available Online at: Retrieved November 10, 2008, from http://www.idcol.org/energyProject.php

69. IEA (2000). Experience Curve for Energy Technology Policy. IEA/OECD, Paris.

70. IEA (2007). Renewable for Heating and Cooling: Untapped Potential. Paris: International Energy Agency, OECD/IEA. Retrieved June 2, 2008, from http://www.iea.

org/textbase/nppdf/free/2007/Renewable_Heating_Cooling_Final_WEB.pdf, accessed on June 2, 2008

71. IEA. (2006a). Barriers to Technology Diffusion: The Case of Solar Thermal Technologies. Paris: OECD/IEA.

72. IEA. (2006b). World Energy Outlook 2006. Paris, France: IEA.

73. IEA, (2009). International Energy Agency (IEA) Database Vol. 2010, Release 01: (a) Energy Balances of Non-OECD Member Countries; (b) Energy Balances of OECD Member Countries. Available at: <http://caliban.sourceoecd.org/vl=213511/cl=25/nw=1/rpsv/iea_database.htm>.

74. IEA, (2011). 2009 Amendment of the Renewable Energy Sources Act – EEG. Global Renewable Energy, Policies and Measures. Available at:

75. <http:// www.iea.org/textbase/pm/?mode=re&id=4054&action=detail>.

76. IEA-PVPS (2007). Trends in Photovoltaic Applications: Survey report of selected IEA countries between 1992 and 2006. IEA Photovoltaic Power Systems Program.

77. IEA-PVPS. (2009). Trends in photovoltaic applications: Survey report of selected IEA countries between 1992 and 2008. International Energy Agency. Photovoltaic Power Systems Programme.

78. Johansson, T. B., McCormick, K., Neij, L., and Turkenburg, W. (2004). The Potentials of Renewable Energy: Thematic Background Paper. Thematic Paper prepared for the International Conference on Renewable Energies, Bonn. Retrieved June 6, 2008, from http://www.iiiee.lu.se/C1256B88002B16EB/$webAll/02DAE4E619978 3A9C1256E29004E1250?OpenDocument; retrieved on July 6th 2008.

79. Kammen, D. M., and Pacca, S. (2004). Assessing the Cost of Electricity. Annual Review of Environment and Resources, 29, 301-344.

80. Kannan, R., Leong, K. C., Osman, R., Ho, H. K., and Tso, C. P. (2006). Life cycle assessment study of solar PV systems: An example of a 2.7 kWp distributed solar PV system in Singapore. Solar Energy, 80(5), 555-563.

81. Klein, A., Held, A., Ragwitz, M., Resch, G., and Faber, T. (2007). Evaluation of Different Feed-in Tariff Design Options: Best practice paper for the International Feed-in Cooperation. Karlsruhe, Germany and Laxenburg, Austria: Fraunhofer Institut für Systemtechnik und Innovationsforschung and Vienna University of Technology Energy Economics Group.

82. Lazard (2009). Levelized Cost of Energy Analysis - Version 3.0. New York: Lazard.

83. Letendre, S., Denholm, P., and Lilienthal, P. (2006). New Load, or New Resource? The Industry Must Join a Growing Chorus in Calling for New Technology. Public Utilities Fortnightly, 144(12), 28-37.

84. Li, J., and Hu, R. (2005). Solar Thermal in China: Overview and Perspectives of the Chinese Solar Thermal Market. Refocus, 6(5), 25-27.

85. Lotker, Michael. 1991. Barriers to Commercialization of Large-Scale Solar Electricity: Lessons Learned from the LUZ Experience. Sandia National Laboratories.

86. Lushetsky, J. (2008). Solar Energy Technologies Program: Accelerating the Future of Solar. Proceedings of the Technology Commercialization Showcase, Washington, DC.

87. Margolis, R. and Zuboy, J. (2006). Nontechnical Barriers to Solar Energy Use: Review of Recent Literature. Technical Report NREL/TP-520-40116, September 2006. Retrieved September 24, 2008, from www.nrel.gov/docs/fy07osti/40116.pdf

88. Mariyappan, J., Bhardwaj, N., Coninck, H., and Linden, N. (2007). A Guide to Bundling Small-scale CDM Projects. Report. Energy Research Center of the Netherlands (ECN). Retrieved November 1, 2008, from http://ecn.nl.

89. Martinot, E., and Li, J. (2007). Powering China's Development: The Role of Renewable Energy. Washington DC: Worldwatch Institute.

90. Mendonça, M. and and D. Jacobs (2009), Feed-in Tariffs Go Global: Policy in Practice, Renewable Energy World Magazine, Vol. 12, No. 4, pp.

91. MNRE (2006). XIth Plan Proposal for New and Renewable Energy, Ministry of New and Renewable Energy, Government of India.

92. Narayanaswami, S. (2001), Interview in The Solar Cooking Archives. Retrieved June 10, 2008, from http://solarcooking.org/

93. NEA (1986). Projected Cost of Generating Electricity from Nuclear and Coal-fired Power Stations for Commissioning in 1995. NEA/IEA, Paris.

94. NEA/IEA (1989). Projected Cost of Generating Electricity from Nuclear and Coal-fired Power Stations for Commissioning in 1995-2000. NEA/IEA, Paris.

95. NEA/IEA (1992). Projected Cost of Generating Electricity- Update 1992. NEA/IEA, Paris.

96. NEA/IEA (1998). Projected Cost of Generating Electricity- Update 1998. NEA/IEA, Paris.

97. NEA/IEA (2005). Projected Cost of Generating Electricity- 2005 Update. NEA/IEA, Paris.

98. NEA/IEA (2007a). Key Energy Statistics 2007. IEA: Paris. Retrieved July 4, 2008, from http://www.iea.org/textbase/nppdf/free/2007/key_stats_2007.pdf

99. NEA/IEA (2007b). IEA Statistics: Electricity Information. IEA: Paris.

100. NEA/IEA (2010). Projected Cost of Generating Electricity- 2010 Edition. IEA: Paris.

101. Neij, L. (2008). Cost development of future technologies for power generation – A study based on experience curve and complementary bottom-up assessments. Energy Policy, 36(6), 2200-2211.

102. O''Rourke, S., Kim, P., & Polavarapu, H. (2009). Solar Photovoltaic Industry: Looking Through the Storm. A Report by Global Markets Research, Deutsch Bank Securities Inc.

103. O'Brien, C., and Rawlings, L. (2006). A Description of An Auction-Set Pricing, Standard Contract Model With 5-Year SREC Generation. In White paper series: New Jersey's solar market (pp. 36-53). Trenton, NJ: New Jersey Clean Energy Program.

104. OECD. (2003). Technology, Innovation, Development and Diffusion. Information Paper. Environment Directorate. Paris: OECD/IEA.

105. Photon: Rogol, M., Flynn, H., Porter, C., Rogol, J., & Song, J-K. (2007). Solar Annual 2007: Big Things in a Small Package. Aachen, Germany: Solar Verlag GmbH/ PHOTON Consulting. Retrieved March 26, 2009, from http://www.photonconsulting.com/solar_annual_2007.phpPodewils, C. (2007, May). The €150 Billion Coup: Funding for PV in Germany has taken on daunting dimensions. PHOTON International, 62-65.

106. Prometheus Institute (PI) (2006). US Solar Industry Year in Review: US solar energy industry changing ahead. Retrieved September 23, 2008, from http://www.

prometheus.org/research/year_in_review_2006 and http://www.prometheus.org/re-search/polysilicon2006

107. PVRES (2007). Large-scale Photovoltaic Power Plants. Cumulative and Annual In-stalled Power Output Capacity. Revised Edition, April 2008. Available online at: http://www.pvresources.com/download/AnnualReport2007.pdf, accessed on June 2, 2008.

108. Radulovic, V. (2001). Are New Institutional Economics Enough? Promoting Photo-voltaics in India"s Agriculture Sector. Energy Policy, 33, 1883-1899.

109. REN21 (2005 to 2011 Issues). Global Status Report. Paris: REN21 Secretariat.

110. Reynolds, A. W. (1983). Projected costs of electricity from nuclear and coal-fired power plants (Vol. 1-2). Washington DC: Energy Information Agency (EIA).

111. Rickerson, W., Ettenson, L., and Case, T. (2007). New York City's Solar Energy Fu-ture: Part II-Solar Energy Policies and Barriers in New York City. New York: Center for Sustainable Energy, Bronx Community College.

112. Rio, P.D. (2007). Encouraging the implementation of small renewable electricity CDM projects: An economic analysis of different options. Renewable and Sustain-able Energy Reviews, 11(7), 1361-1387.

113. Roethle Group (2002). The Business of Fuel Cells: What's Happening with Fuel Cells?

114. SELCO (2009). Access to Sustainable Energy Services via Innovative Financing: 7 Case Studies. Bangalore: SELCO Solar Light Pvt. Ltd. Retrieved April 20, 2009, from http://www.selco-india.com/pdfs/selco_booklet_web.pdf

115. Shell (2008). Shell Energy Scenario to 2050. Retrieved November 29, 2008 from http://shell.com/scenarios

116. Stoddard, L. Abiecunas, J., and O'Connell, R. (2007). Economic, Energy, and En-vironmental Benefits of Concentrating Solar Power in California. National Renew-able Energy Labaratory (NREL): Golden, ColoradoSunPower (2008). Residential Panels for Homes. Retrieved April 20, 2009, from http://www.sunpowercorp.com/Products-and-Services/Residential-Solar-Panels.aspx

117. Synapse Energy Economics Inc.(SEEI). (2005). Feasibility Study of Alter-tive En-ergy and Advanced Energy Efficiency Technologies for Low-Income Housing in Massachusetts. Cambridge, MA: SEEI.

118. Teske, S., Zervos, A., and Schäfer, O. (2007). Energy Revolution: A Sustainable World Energy Outlook: Greenpeace International and European Renewable Energy Council.

119. Turkenburg, W. C. (2000). Renewable energy technologies (Ch.7). In: World Energy Assessment (WEA). UNDP, New York.

120. Uddin, S. K., and Taplin, R. (2008). Toward Sustainable Energy Development in Bangladesh. The Journal of Energy and Development, 17 (3), 292-315.

121. UNEP Risoe Center. (2011). CDM/JI Pipeline Analysis and Database, 2011. Re-trieved July 18, 2011, from http://www.cdmpipeline.org.

122. United Nations Framework on Climate Change Convention (UNFCCC). (2008). CDM Statistics. Retrieved November 3, 2008 from http://www.unfccc.org.

123. USPVIR. (2001). Solar Electric Power: The U.S. Photovoltaic Industry Roadmap(USPVIR) Reprint of 2001. Retrieved September 23, 2008, from www.sandia.gov/pv/docs/PDF/PV_Road_Map.pdf

124. Waegel, A.; Byrne, J.; Tobin, D.; and Haney, B. 2006a."Hydrogen Highways: Lessons on the Energy Technology-Policy Interface." Bulletin of Science, Technology and Society, Vol. 26, No. 4. Pp. 288-298.

125. Weiss, W., Bergmann, I., and Stelzer, R. (2009). Solar Heat Worldwide: Markets and Contribution to the Energy Supply 2007. Paris: International Energy Agency Solar Heating and Cooling Program

126. Weiss, W. , Bergmann, I., and Faninger, G. (2005 to 2011 Issues). Solar Heat Worldwide: Markets and Contribution to the Energy Supply 2006. Paris: International Energy Agency Solar Heating and Cooling Program

127. Wenzel, E. (2008). Barriers to Solar Energy's Blockbuster Promise. Retrieved September 23, 2008, from http://news.cnet.com/8301-11128_3-9939715-54.html

128. Willis, M., Wilder, M., Curnow, P., and Mckenzie, B. (2006). The Clean Development Mechanism: Special Considerations for Renewable Energy Projects. Renewable Energy and International Law Project. Retrieved September 20, 2008 from, http://www.reilproject.org/.

129. Winter, C. J., Sizmann, R. L., and Vant-Hull, L. L. (1991). Solar Power Plants. Springer-Verlag Berlin and Heidelberg GmbH & Co. K

130. Wolff, G., Gallego, B., Tisdale, R., and Hopwood, D. (2008). CSP Concentrates the Mind. Renewable Energy Focus, January/February, 42-47.

CHAPTER 8

ECONOMICAL EVALUATION OF LARGE-SCALE PHOTOVOLTAIC SYSTEMS USING UNIVERSAL GENERATING FUNCTION TECHNIQUES

YI DING , WEIXIANG SHEN, GREGORY LEVITIN, PENG WANG, LALIT GOEL, AND QIUWEI WU

8.1 INTRODUCTION

With the ever increasing concerns on environmental issues and the depletion of fossil fuels, the photovoltaic (PV) technology has drawn great attention and remarkable investments in the past decade [1]. This is due to the fact that the PV technology shows many advantages over other renewable energy technologies in terms of modularity, expandability, maintenance and reliability. In recent years, the contribution of the PV power generation to the grid has been rapidly increasing; at the current growth rate, it is expected to reach 2% of the world electricity generation by 2020 and up to 5% by 2030 [2, 3]. During the next ten years, up to 15% of electricity in European Union will be produced by solar energy resources [4].

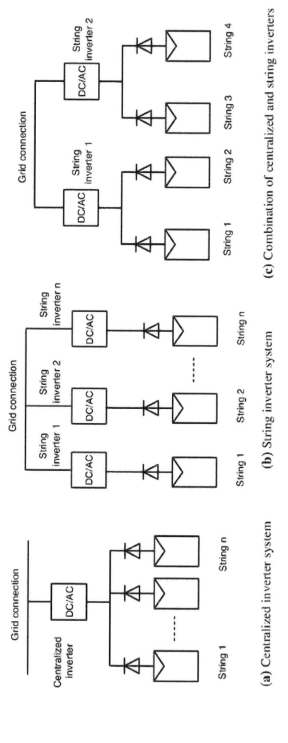

(a) Centralized inverter system

(b) String inverter system

(c) Combination of centralized and string inverters

FIGURE 1: Configurations of PV system

An important question for investors, planners and regulators is the return and cost of a PV project. The cost structure of PV systems is different from that of conventional generation system using fossil fuels such as coal, oil or natural gas. The initial capital cost is higher: basic components of a PV system—solar panels are quite expensive. However prices of solar panels are dropping fast: the average one-off installation cost of solar panels has already dropped from more than $2 per unit of generating capacity in 2009 to about $1.50 in 2011 [5]. On the other hand, there are no fuel cost and greenhouse gas emissions during the lifespan operation of 20–30 years. The maintenance cost of PV system is also relatively low.

PV systems are complicated engineering systems. A PV system is mainly composed of many solar panels and DC/AC inverters. The trend of the fast growing PV systems is to adopt large-scale PV systems, which may require tens or hundreds of solar panels. Depending on input voltage ranges, maximum input currents and capacities of inverters, several solar panels are connected in series to form a string and a few strings are paralleled and tied up to a centralized-inverter or each of the strings is directly interfaced by a separate string-inverter or a combination of both, which are illustrated in Fig. 1. Different configurations have their own performance efficiencies for electricity production. When the performances of those configurations of PV systems are evaluated, it is assumed that the systems work without interruption. Although PV systems are relatively reliable, they may fail occasionally. Ignoring the effects of those failures may result in an optimistic estimation of energy production, which also decreases accuracy of cost assessment.

The approaches for improving the engineering system reliabilities are to increase the redundancy or/and reliability of the components in the system. For example, the use of multiple inverters in PV systems can increase system reliabilities. These approaches can improve the reliability of the PV systems and hence its energy production, but they may result in higher system cost. The reliability based cost assessment for renewable energy systems (RESs) and restructured generation systems has been studied in some recent research. Reference [6] provided a comprehensive analysis of the reliability and its cost implications on various choices of installation sites and operating policies as well as energy types, sizes and mixes in capacity expansion of the RESs. The genetic algorithm was used to optimize the offshore wind farms considering both energy production cost and system reliability [7]. A framework for analyzing the adequacy uncertainties of distributed generation systems was proposed in [8]. However reliability based economical evaluation of large-scale PV systems has not been comprehensively studied, which may be a useful analytical tool for assisting stakeholders in making optimal decision.

The large-scale PV system can be modeled as a typical multi state system (MSS). The UGF technique provides a systematic method for the performance and reliability assessment of MSS, which can replace extremely complicated combinational algorithms and reduce the computational burden [9–11]. Moreover, the UGF technique provides a flexible approach for representing reliability models of various energy systems. The UGF technique and genetic algorithm were used to determine the optimal structure of power systems subject to reliability constraints [12]. The reliability of flow transmission system was analyzed by using the combination of the UGF technique and extended block diagram methods [13]. The redundancy analysis of inter-connected generating systems was discussed in [14]. In [15], the UGF technique was used to determine the reserve expansion for maintaining the reliability level of power systems with high wind power penetration.

In this paper, the UGFs representing probabilistic performance distributions of solar panel arrays, PV inverters and energy production units (EPUs) are developed. The expected energy production models for PV systems under different configurations are also developed. The life cycle cost and annualized life cycle cost are evaluated to conduct economical assessment of a PV project. Moreover, a new economical index for PV systems—expected unit cost of electricity (EUCE) is developed for providing useful information.

Section 2 presents reliability models of large-scale PV systems. The developed UGFs are used to evaluate expected energy production. Cost analysis of PV systems is conducted in Section 3. Section 4 proposes a methodology to identify the feasible configurations of PV systems and determine the optimal one at the minimum EUCE. The proposed methods are used to assess the PV system configurations for providing electricity for a water purification process in Section 5. The conclusions are summarized in Section 6.

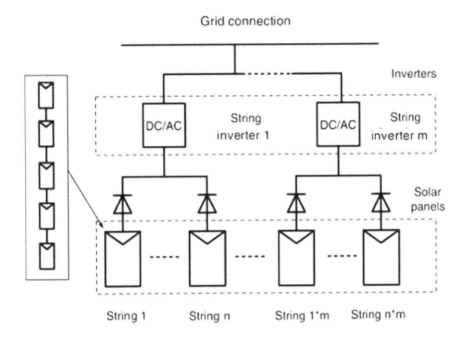

FIGURE 2: Generalized configuration of PV system

8.2 RELIABILITY MODELS OF PV SYSTEMS

A large-scale PV system basically consists of two major parts: solar panels and DC/AC inverters. Figure 2 shows a generalized configuration of the PV system. In the following, the reliability models of solar panel arrays, PV inverters and EPUs, and expected energy production calculation are discussed.

8.2.1 RELIABILITY MODELS FOR SOLAR PANEL ARRAYS

Solar panels are the key components of the PV systems. Solar panels can fail due to the degradation of mechanical properties of encapsulants, the adhesional strength, the presence of impurities, metalization, solder bond integrity and breakage, corrosion, and aging of backing layers, etc.

Given the failure rate λ_i and repair rate μ_i of the solar panel i, the corresponding availability A_i can be calculated by

$$A_i = \mu_i / (\lambda_i + \mu_i) \tag{1}$$

where λ_i is the failure rate referring to the rate of departure from a component up-state (successful state) to its down-state (failure state) and μ_i is the repair rate referring to the rate of departure from the down-state to the up-state.

Some strings consist of several solar panels and a blocking diode in series. Any failure of a solar panel or a diode leads to the total failure of the string. Therefore, the availability of the string s can be evaluated by:

$$A_s = \prod_{i=1}^{n} A_s \cdot A_d \tag{2}$$

where n is a number of solar panels in the string and A_d is the availability of the diode:

$$Ad = \mu d / (\lambda d + \mu d) \tag{3}$$

where λ_d and μ_d are the failure rate and the repair rate of the diode, respectively.

The available capacity W s of the string s can be calculated as:

$$W_s = \sum_{i=1}^{n} W_i - W_d \tag{4}$$

where W_i is the available capacity of the solar panel i; W_d is the capacity loss caused by the blocking diode, which can be determined as

$$W_d = U_d \cdot I_s \tag{5}$$

where U_d is the voltage drop across the blocking diode and I_s is the current of the string.

The UGF technique is proved to be very convenient for numerical realization and requires small computational resources [9–11] for performance and reliability evaluation of engineering systems [9]. Therefore, the UGF technique is used to evaluate the expected energy production of the PV system. The UGF representing the capacity distribution of a string s can be defined as a polynomial:

$$U_s(Z) = \sum_{k_s=1}^{2} p_{s,k_s} \cdot Z^{W_{s,k,s}} \tag{6}$$

where $p_{s,ks}$ and $w_{s,ks}$ are the probability and the capacity level of state k_s for the string s, $U_s(Z)$ represents the capacity distribution of the string s, Z represents the Z-transform of any discrete random variables that has the probability mass function taking the form shown in (6) [10].

The string s has two states: failure state and successful state. For the failure state, the capacity level and unavailability are 0 and $(1-A_s)$, respectively. For the working state, the capacity level and availability are W_s and A_s, respectively.

A few strings are also arranged in parallel to form solar array and connected to a string inverter. Failure of any string in the array is tolerated without the loss of an entire array. However, the failure of a string degrades the available capacity of the array, leading to several de-rated states of the array. As a result, the solar array in the PV system can be regarded as a MSS. The parallel operator $\Omega_{\phi p}$ is applied for the parallel MSS by using associative and commutative properties. The parallel operator is a kind of composition operator to calculate the UGF for the parallel MSS, which strictly depends on the properties of the parallel structure function [9, 10]. For example, if elements are connected in parallel, its capacity level for the state k_s is the sum of the corresponding capacities $w_{s,ks}$ ($s = 1, 2,..., N$) of its elements, and the structure function for such a subsystem takes the form:

$$\phi_p(w_{1,k_1}, w_{1,k_2}, ..., w_{N,k_N}) = \sum_{s=1}^{N} w_{s,k_s} \tag{7}$$

The capacities of elements unambiguously determine the capacity of the subsystem or system. The transform, which maps the space of the element capacities into the space of the system capacity, is the system structure function [9, 10].

For a solar array with N strings in parallel, its UGF can be obtained based on the UGFs for the arrays using the parallel composition operator $\Omega_{\phi p}$ over UGF representations of N strings:

$$U_a(Z) = \sum_{k_a=1}^{2^N} P_{a,k_a} \cdot Z^{W_{a,ka}}$$

$$= \Omega_{\phi p} \left(\sum_{k_1=1}^{2} P_{1,k_1} \cdot Z^{W_{1,k1}}, P_{2,k_2} \cdot Z^{W_{2,k2}}, ..., \sum_{k_N=1}^{2} P_{N,k_N} \cdot Z^{W_{N,kN}} \right)$$

$$= \sum_{k_1=1}^{2} \sum_{k_2=1}^{2} ... \sum_{k_N=1}^{2} \left(\prod_{s=1}^{N} P_{s,k_s} \cdot Z^{\sum_{s=1}^{N} w_{s,ks}} \right) \tag{8}$$

where $p_{a,ka}$ and $w_{a,ka}$ are the probability and the available capacity of the array in the state k_a, respectively. Equation (8) represents the capacity distribution of the solar array [8]: the coefficients of the terms in the polyno-

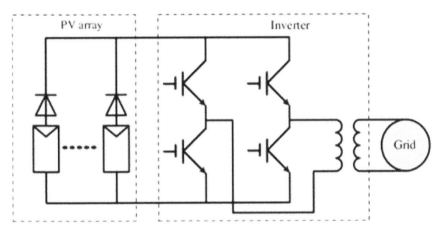

FIGURE 3: Structure of inverter

mial (8) represent the probabilities of the array states while the exponents represent the corresponding capacities. The array has 2 N states.

8.2.2 RELIABILITY MODELS FOR INVERTERS

PV inverters convert DC power from solar array into AC power, which matches the voltage of power grids. It is believed that inverters are the reliability bottleneck of PV systems and the vast majority of PV system failures are caused by inverters.

A typical PV inverter as shown in Fig. 3 includes four insulated-gate bipolar transistors (IGBTs) and an isolation transformer.

The failure rate of an IGBT is affected by the operating environment and other factors [16], which can be evaluated by

$$\lambda_{IGBT} = 0.5 \cdot \lambda_b \cdot \pi_T \cdot \pi_Q \cdot \pi_E \tag{9}$$

where λ_b is the base failure rate, π_T, π_Q and π_E are the temperature factor, the quality factor and the environment factor, respectively. Therefore, the availability of the IGBT can be calculated as

$$A_{IGBT} = \mu_{IGBT} / (\lambda_{IGBT} + \mu_{IGBT}) \tag{10}$$

where μ_{IGBT} is the repair rate of the IGBT.

Similar equations can also be used to evaluate the availability A_{Trans} of an isolation transformer. The components in a PV inverter are modeled as functional blocks connected in series. Any failure of an IGBT or an isolation transformer leads to the failure of an inverter. As a result, the availability of the PV inverter can be evaluated by

$$A_I = (A_{IGBT})^m \cdot A_{Trans} \tag{11}$$

where m is the number of IGBTs in the inverter. Therefore, the UGF representing the capacity distribution of the PV inverter is defined as

$$U_I(Z) = \sum_{k_i=1}^{2} P_{I,k_i} \cdot Z^{W_{i,ki}} \tag{12}$$

where $p_{I,kI}$ and $w_{I,kI}$ are the probability and the capacity of state k_I for the inverter.

The PV inverter has two states: failure state and successful state. For the failure state, the capacity level and the unavailability are 0 and $(1-A_I)$, respectively. For the successful state, the availability is A_I, and the capacity is determined by the nominal capacity of the inverter $w_{I,n}$ and the efficiency of the inverter e_I:

$$W_{I,2I} = W_{I,n} \cdot e_I \tag{13}$$

8.2.3 RELIABILITY MODELS FOR ENERGY PRODUCTION UNIT

A PV inverter is connected in series with the solar array to form an EPU. The series operator $\Omega_{\phi s}$ is used to calculate the UGF for an EPU using associative and commutative properties. For the type of the MSS containing elements connected in series, its capacity level for the state k_E is the minimization of the corresponding capacities of its components. The structure function for an EPU takes the form:

$$\phi_s(w_{I,kI}, w_{a,ka}) = \min(w_{I,kI}, w_{a,ka}) \tag{14}$$

The capacity distribution of the EPU E can be obtained based on the UGF representing capacity distribution of the array and the PV inverter by using the series composition operator $\Omega_{\phi s}$:

$$U_E(Z) = \sum_{k_E=1}^{N_E} P_{E,k_E} \cdot Z^{W_{E,kE}} = \Omega_{\phi s} \left(\sum_{k_a=1}^{2^N} P_{a,k_a} \cdot Z^{W_{a,ka}}, \sum_{k_I=1}^{2} P_{I,k_I} \cdot Z^{W_{I,kI}} \right)$$

$$= \sum_{k_a=1}^{2^N} \sum_{k_1=1}^{2} \left(P_{a,k_a} \cdot P_{I,k_I} \cdot Z^{\min(w_{a,ka}, w_{I,kI})} \right) \tag{15}$$

where N_E is the number of states of the EPU and equals to $2(N+1)$.

8.2.4 EXPECTED ENERGY PRODUCTION

The PV system consists of several EPUs in parallel to supply electricity to the power grid. With the UGF for each EPU, the UGF representing the capacity distribution for the entire PV system can be calculated by using the parallel composition operator $\Omega_{\phi p}$:

$$
U_{sys}(Z) = \sum_{k=1}^{M} P_k \cdot Z^{W_k}
$$

$$
= \Omega_{\phi p} \left(\sum_{k_1=1}^{N_1} P_{1,k_1} \cdot Z^{W_{1,k_1}}, \sum_{k_2=1}^{N_2} P_{2,k_2} \cdot Z^{W_{2,k_2}}, \ldots, \sum_{k_E=1}^{N_E} P_{E,k_E} \right.
$$

$$
\left. \cdot Z^{W_{E,k_E}}, \ldots, \sum_{k_m=1}^{N_m} P_{m,k_m} \cdot Z^{W_{m,km}} \right)
$$

$$
= \sum_{k_1=1}^{N_1} \sum_{k_2=1}^{N_2} \ldots \sum_{k_m=1}^{N_m} \left(\prod_{E=1}^{m} P_{E,k_E} \cdot Z^{\sum_{E=1}^{m} E_{E,k_E}} \right)
$$

(16)

where m, M, p_k and w_k are the number of EPUs, the state number of the PV system and the probability, and capacity level of state k for the PV system, respectively.

The general technique for determining the UGF of the PV system is based on a state enumeration approach. This approach is usually extremely resource consuming. Fortunately, the PV system can be divided into subsystems (string, array and EPU) and the UGF method allows one to obtain the system UGF recursively. This property of the UGF method is based on the associative property of many practically used structure functions. The recursive approach presumes the UGF of subsystems containing several basic components and then treating the subsystem as a single component with the obtained UGF when the UGF of a higher level subsystem is computed [9]. The recursive approach provides a drastic reduction of the computational resources needed to obtain the capacity distribution of a complex MSS.

The yearly expected energy production (YEEP) of the PV system EE_{pv} is defined as the product of the expected capacity of the system Ew and yearly peak sun hours PSH:

$$EE_{pv} = E_w \times PSH \qquad (17)$$

The operator δ_w is used to calculate Ew and defined as

$$E_w = \delta_w\left(U_{sys}\right) = \delta_w\left(\sum_{k=1}^{M} P_k \cdot Z^{W_k}\right) = \sum_{k=1}^{M} p_k \cdot w_k \qquad (18)$$

where U_{sys} is obtained from (16).

8.3 COST ANALYSES OF PV SYSTEMS

The cost of the PV system includes acquisition cost, operating and maintenance cost. These costs can be divided into two types. One is the recurring cost, e.g., operation and maintenance (O&M) cost. The other is the initial capital cost, e.g., investment cost for purchasing solar panels and inverters. The life cycle cost (LCC) analysis evaluates the total system cost during the life span of the system. The LCC for all the parts in the system is added together to obtain the LCC for the entire system, where the system life cycle is assumed to be T years. The LCC analysis converts the recurring cost into the present worth [17]. Annualized life cycle cost (ALCC) is also evaluated to provide an annualized "payment" required to fund the total system cost over the life span [18]. However LCC and ALCC analysis cannot evaluate the "equivalent" unit cost for producing electricity, which is important for determining the most cost-efficient system design. Therefore, a new economical index—EUCE is proposed to provide an informative metrics for evaluating cost efficiency of PV systems. The EUCE is defined as the system ALCC divided by YEEP.

LCC of the system is

$$LCC = C_{solar} + C_{inverter} + C_{om} \tag{19}$$

where C_{solar}, $C_{inverter}$, and C_{om} are the costs of solar, inverter, and operation and maintenance (O&M), respectively,

$$C_{solar} = \text{price of a solar panel} \times \text{number of solar panel} \tag{20}$$

$$C_{inverter} = \text{price of an inverter} \times \text{number of inverters} \tag{21}$$

$$C_{om} = C_{om0} \cdot P_{a1} \tag{22}$$

$$P_{a1} = X \cdot P_{a} \tag{23}$$

$$P_{a} = (1 - X^{T}) / (1 - X) \tag{24}$$

$$X = (1 + i) / (1 + d) \tag{25}$$

where P_{a} is the present worth, X_{T} is the present worth factor for a cost in T years, i is the inflation rate, and d is the interest rate [17].

Annualized life cycle cost (ALCC) is

$$ALCC = LCC / P_{a} \tag{26}$$

Expected unit cost of electricity (EUCE) is

$$EUCE = ALCC / EE_{pv} \qquad (27)$$

8.4 FEASIBLE CONFIGURATION IDENTIFICATION FOR PV SYSTEMS

As shown in Fig. 2, a number of solar panels are connected in series to form a string, and a number of strings are paralleled and connected to a string inverter, then all the string inverters are connected to the power grid. Thus, in principle there are enormous configurations of the PV system for the given large number of solar panels. However, the feasible configurations are practically constrained by input voltage ranges, maximum input currents and capacities of the inverters. Consequently, only those configurations whose voltages, currents and capacities are within the normal operation range of the inverters will be considered. The constraints are specified as follows:

1. Input voltage limits

   $$V_I^{min} \le V_s \le V_I^{max} \qquad (28)$$

   where V_s is the voltage of the string connected to a string inverter, if the string consists of n solar panels in series, then $V_s = n \cdot V_p$, V_p is the voltage of the solar panel; V_I^{min} and V_I^{max} are the lower and upper limits, respectively.

2. Input current limit

   $$I_I \le I_I^{max} \qquad (29)$$

   where I_I is the input current of the string inverter, I_I^{max} is the maximum input current. If the array consists of N strings in parallel, then $I_I = N \cdot I_s$, where I_s is the current of the string.

3. Capacity limit

$$W_a \leq W_I \tag{30}$$

where W_a is the available capacity of a solar array tied up to a string inverter, W_I is the nominal capacity of the string inverter. $W_a = N \cdot n \cdot W_p$, where W_p is the available capacity of the solar panel in the string. This constraint indicates that the available capacity of solar array tied up to a string inverter should be less than or equal to the nominal capacity of the inverter.

For various feasible configurations of the PV system, the economical efficiency is the major concern for comparing different design options. The reliability based cost analysis allows the investor or designer to evalu-

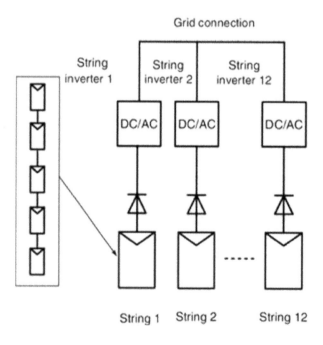

FIGURE 4: Twelve string configurations with 12 inverters

ate the effects of different design options. A EUCE analysis discussed in the previous section can be helpful for determining the most cost-efficient system configuration.

The following three steps can be implemented for determining the most economically efficient design option:

1. Identifying all feasible configurations satisfying constraints (28)–(30). These identified configurations include the connections of solar panels, namely the number of solar panels in series and the number of strings in parallel, and the number of inverters required.
2. Evaluating the expected energy production and the EUCE of each configuration.
3. Determining the optimal configuration of the PV system at the minimum EUCE obtained from step 2).

8.5 APPLICATION

The proposed method is used to assess the performance and determine the feasible configurations of the PV system, which provides energy for a national demonstration project of water purification process in Singapore. According to electrical energy requirement of the process, it is estimated that 60 pieces of 175 Wp (peak power) solar panels from SolarWorld are needed. For these 60 solar panels, all feasible configurations of the system are identified based on the input voltage ranges, the maximum input currents and the nominal capacities of the inverters commercially available in the market, as shown in Table 1 and Figs. 4–7. Figure 4 shows one configuration of the system which consists of 12 strings each having 5 panels in series, Fig. 5 illustrates one configuration which consists of 10 strings each having 6 panels in series, Fig. 6 shows three configurations of the systems which consist of 6 strings each having 10 panels in series, and Fig. 7 shows one configuration of the system which consists of 5 strings each having 12 panels in series. It should be noted that the special "text string" is defined to represent each configuration, for example, the "c12p12s05" in Fig. 4 represents the configuration that includes 12 PV inverters, 12 strings and each having 5 solar panels in series.

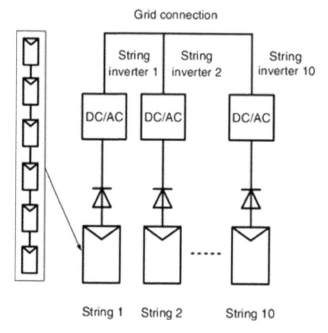

FIGURE 5: Ten string configurations with 10 inverters

TABLE 1: Parameters of inverters for evaluation of YEEP and system cost

Configuration	Capacity (W)	Voltage (V)	Current (A)	Prices (S$)	O & M (S$)	Peak efficiency (%)
c12p12s05	1,100	139–400	10	2,559	256	93.0
c10p10s06	1,100	139–400	10	2,559	256	93.0
c06p06s10	2,500	224–600	12	4,444	444	94.1
c03p06s10	3,800	200–500	20	5,083	508	95.6
c02p06s10	5,500	246–600	26	8,156	816	96.1
c05p05s12	2,500	224–600	12	4,444	444	94.1

Note: one solar panel cost is S$1,400 and its monthly O&M cost is assumed to 1% of solar panel cost, namely S$14 each panel

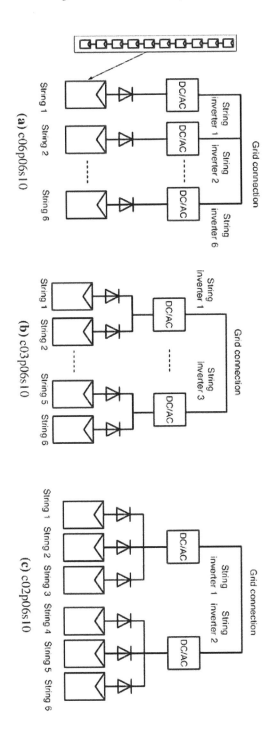

FIGURE 6: Six string configurations with different numbers of inverters

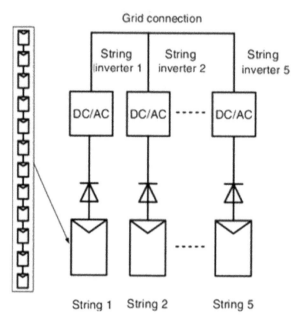

FIGURE 7: Five string configuration with 5 inverters

The basic parameters of the PV inverters including nominal capacity, input voltage range, purchasing cost, O&M cost and peak efficiency for different configurations are shown in Table 1 [19].

For evaluating the failure rate of the IGBT of the inverter shown in Fig. 3, the base failure rate λ_b is set as 0.060, the temperature factor π_T is computed from $\pi T = \exp[-1{,}925([1/(T_j+273)] \cdot 1/298)]$ [20], where T_j is the junction operating temperature of the device and set as 40°C, $\pi_Q = 5.0$ and $\pi_E = 1.0$ for the other conditions.

The repair rate of the IGBT equals to 0.0017 per hour. The failure rates of the solar panel and the string diode are set as 0.2068 and 0.0198 per million hours, respectively; the repair rates of the solar panel and the string diode are 4.0556 per year [20]. The current of the solar panel string is 4.89 A at the maximum power and the voltage drop of the blocking diode is set as 2.0 V. The yearly peak sun hours (PSHs) equals to 1,721.7 hours in Sin-

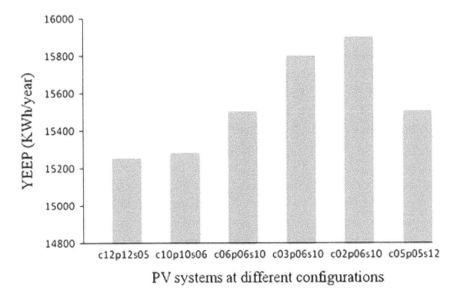

FIGURE 8: Yearly expected energy production of PV systems

gapore, which is the average yearly PSHs over the period of 1993 to 2007 [21]. The inflation rate and the interest rate are assumed as 2.1% and 1%, respectively. The system life cycle is assumed to be 20 years.

With all these parameters, the YEEP for each configuration is evaluated and the results are shown in Fig. 8. It can be observed from Fig. 8 that the configuration of "c02p06s10" has the highest expected energy production (15,898.6 kWh/year), which consists of 2 inverters, 6 strings, each of the strings having 10 solar panels in series. The YEEP for different configurations ranges from 15,255.7 kWh/year to 15,898.6 kWh/year, with the difference of 4.21%. It represents that the difference of expected energy production for different configurations in the system life cycle can be 12,860 kWh.

The differences in the YEEP are mainly caused by the reliability differences of PV arrays and the connected PV inverters for various configurations. The assessment of the ALCC for each configuration of the PV systems is also conducted. Consequently, the EUCE is easily calculated as the ratio of the ALCC to the YEEP, as shown in Fig. 9.

It can be seen that the EUCE for different configurations ranges from 0.434–0.598 S$/kWh. The configuration of "c03p06s10" has the lowest EUCE (0.434 S$/kWh), which also has the second highest YEEP. The configuration of "c02p06s10" has the second lowest EUCE (0.441 S$/kWh) with the highest YEEP. The configuration of "c12p12s05" has the highest EUCE (0.598 S$/kWh). Therefore, the "c03p06s10" is the optimal configuration with the lowest EUCE. The comparison results show that simply increasing the system cost by using relatively large number of low capacity inverters in the PV systems cannot guarantee high expected energy production, instead the configuration with high reliable PV arrays, and high capacity inverters can achieve the lowest unit cost of electricity.

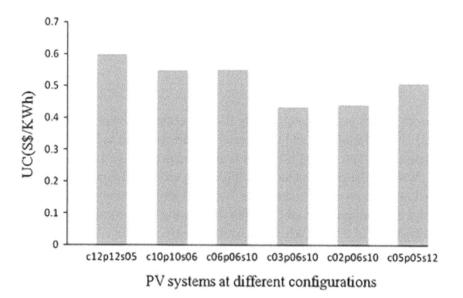

FIGURE 9: Unit cost of PV systems

8.6 CONCLUSION

In this paper, the UGF technique is used to represent reliability models of solar panel arrays, PV inverters and EPUs in a large-scale PV system. Based on the developed probabilistic performance distribution models, the expected energy production for PV systems is evaluated with respect to the reliability of system elements. The reliability based cost analysis of PV systems is conducted for providing informative metrics to stakeholders for making the optimal decision. A new economical index for PV systems—EUCE is also developed in this paper. The proposed method is used to identify the feasible configurations of PV systems and determine the economically optimal one.

REFERENCES

1. Hoffmann W (2006) PV on the way from a few lead markets to a world market. In: Conference Record of the IEEE 4th World Conference on Photovoltaic Energy Conversion, Vol 2, Waikoloa, HI, USA, 7–12 May 2006, pp 2454–2456
2. IEA (2006) Trends in photovoltaic applications: survey report of selected IEA countries between 1992 and 2005. IEA-PVPS T1-15: 2006, The International Energy Agency, Paris, France
3. IEA (2006) Chapter 6: Electricity. In: International Energy Outlook. Energy Information Administration, US Department of Energy, Washington, DC, USA
4. Commission of the European Communities (2009) Communication from the commission to the European parliament, the council, the European Economic and Social Committee and the Committee of the Regions: investing in the development of low carbon technologies (SET-Plan). Commission Staff Working Document, Brussels, Belgium
5. Clark D (2011) Price of solar panels to drop to $1 by 2013. 2011-06-20. http://www.guardian.co.uk/environment/2011/jun/20/solar-panel-price-drop
6. Rajesh K, Billinton R (2001) Reliability/Cost implications of PV and wind energy utilization in small isolated power systems. IEEE Trans Energ Conver 16(4):368–373
7. Zhao M, Chen Z, Blaabjerg F (2009) Optimization of electrical system for offshore wind farms via genetic algorithm. IET Renew Power Gener 3(2):205–216
8. Li YF, Zio E (2012) Uncertainty analysis of the adequacy assessment model of a distributed generation system. Renew Energ 41:235–244
9. Lisnianski A, Levitin G (2003) Multi-state system reliability: assessment, optimization and applications. World Scientific, Singapore

10. Levitin G (2005) Universal generating function and its applications. Springer, Berlin, Germany
11. Ding Y, Lisnianski A (2008) Fuzzy universal generating functions for multi-state system reliability assessment. Fuzzy Set Syst 159(3):307–324
12. Lisnianski A, Levitin G, Ben-Haim H et al (1996) Power system structure optimization subject to reliability constraints. Electr Power Syst Res 39:145–152
13. Lisnianski A (2007) Extended block diagram method for a multi-state system reliability assessment. Reliab Eng Syst Safe 92(12):1601–1607
14. Lisnianski A, Ding Y (2009) Redundancy analysis for repairable multi-state system by using combined stochastic processes methods and universal generating function technique. Reliab Eng Syst Safe 94(11):1788–1795
15. Ding Y, Wang P, Goel L et al (2011) Long term reserve expansion of power systems with high wind power penetration using universal generating function methods. IEEE Trans Power Syst 26(2):766–774
16. Military handbook: reliability prediction of electronic equipment. MIL-HDBK-217F, US Department of Defense, Washington, DC, USA
17. Markvart T (2000) Solar electricity, 2nd edn. John Wiley & Sons Inc, New York, USA
18. Hoff J (2006) Equivalent uniform annual cost: a new approach to roof life cycle analysis. In: Proceedings of the RCI 21st International Convention and Trade Show, Phoenix, AZ, USA, 23–38 Mar 2006, pp 115–124
19. Solarzone. http://www.solazone.com.au/gridinvert.htm
20. Copley controls. http://www.copleycontrols.com/Motion/pdf
21. Solar energy engineering. http://energy.caeds.eng.uml.edu/

CHAPTER 9

SIMULATING THE VALUE OF CONCENTRATING SOLAR POWER WITH THERMAL ENERGY STORAGE IN A PRODUCTION COST MODEL

PAUL DENHOLM AND MARISSA HUMMON

9.1 INTRODUCTION

Concentrating solar power (CSP) becomes a dispatchable source of renewable energy by adding thermal energy storage (TES). There have been a limited number of analyses that examine the value of this energy source and how this value varies as a function of grid configuration and fuel prices.

Challenges of properly valuing CSP include the complicated nature of this technology. Unlike completely dispatchable fossil sources, CSP is a limited energy resource, depending on the hourly and daily supply of solar energy. This supply of energy is both variable and not entirely predictable. This requires the limited energy available to be optimally dispatched to provide maximum value to the grid. The actual dispatch of a CSP plant, including its ability to provide ancillary services, will vary as a function of

This chapter was originally published by the U.S. Department of Energy. U.S. Department of Energy, National Renewable Energy Laboratory, Simulating the Value of Concentrating Solar Power with Thermal Energy Storage in a Production Cost Model, by Paul Denholm and Marissa Hummon, NREL/TP-6A20-56731 (November 2012). http://www.nrel.gov/docs/fy13osti/56731.pdf(accessed 30 June 2014).

generator mix, the penetration of variable generation (VG) sources, such as wind and solar photovoltaics (PV), and the amount of storage deployed with CSP.

The ability to evaluate CSP under multiple scenarios requires the use of detailed grid simulation tools, such as a production cost model. A number of commercial production cost models exist, and these are routinely used by utilities, system operators, and researchers to evaluate the impacts of various generation sources. However, there have been limited studies of CSP with TES in the United States using commercial production cost models. Several studies that included CSP assumed that the dispatch of CSP is fixed and did not evaluate the complete benefits of this dispatchable resource (CAISO 2011; GE 2010). Other studies that have included dispatchable CSP made no attempt to isolate the value proposition for CSP or how that value proposition changes with increased levels of wind and solar generation (US DOE 2012; Mai et al. 2012; Denholm et al. 2012).

To completely identify the benefits of CSP and perform analysis in a framework accepted by utilities and system operators, CSP with TES needs to be incorporated into commercially available software. This document describes the methodology of implementing CSP with and without TES into the PLEXOS production cost model. It also provides a preliminary analysis of CSP with TES in a test system, based on two balancing areas located largely in the State of Colorado. It compares the dispatch of systems with CSP and TES to systems with only variable solar generation and examines several performance metrics, including avoided fuel and total system production cost.

9.2 PRODUCTION COST SIMULATIONS

A number of commercial production cost models are available to utilities, system operators, and planners to evaluate the operation of the grid. These models are used to help plan system expansion, evaluate aspects of system reliability, and estimate fuel costs, emissions, and other factors related to system operation. The models have the primary objective function of committing and dispatching the generator fleet to minimize the total cost of production while maintaining adequate operating reserves to meet contingency events and regulation requirements. Modern production cost models often include transmission power flow simulations to ensure basic transmission adequacy for the generator dispatch. These models are increasingly used to evaluate the impact of incorporating VG sources, such as wind and solar. Integration studies evaluate the impact of VG on power plant ramping and reserve requirements and explore changes to grid operations needed to incorporate increasing amounts of VG (GE 2010).

As the penetration of VG increases, studies have found an important role for grid flexibility techniques and technologies, including new market structures, flexible generators, demand response, and energy storage. One option for flexible renewable generation is CSP with TES. This dispatchable energy source can provide grid flexibility by shifting energy over time, providing ancillary services, and ramping rapidly on demand, enabling a greater penetration of VG sources, such as wind and solar PV (Denholm and Mehos 2011). Several previous studies have included CSP to various degrees. The Western Wind and Solar Integration Study (WWSIS) (GE 2010) included CSP with TES but assumed CSP was dispatched in fixed schedules. Integration studies by the California Independent System Operator have included CSP but assumed very little storage (CAISO 2011). Two more recent studies, the SunShot Vision Study (U.S. DOE 2012) and the Renewable Electricity Futures Study (Mai et al. 2012) incorporated CSP with TES into a commercial production cost model (Brinkman et al. 2012) and allowed the model to dispatch the TES resource. These studies demonstrate qualitatively the value of dispatchable solar but did not attempt to isolate the value of CSP with TES or compare how the value of CSP changes as a function of storage or other grid components. Alternatively, there have been studies that focused on the value of CSP with TES but were limited in modeling resolution. An example is a study that used a "price-taker" approach to dispatch a CSP plant against historic prices, assuming these prices (and solar availability) are known with varying degrees of certainty (Sioshansi and Denholm 2010). This type of study can identify some of the additional value that TES adds in terms of energy shifting and ancillary services; however, the value of this analysis is limited because it cannot examine the impact of different fuel prices, grid mixes, or the ability of CSP to interact with variable renewable sources, such as wind and PV.

A more comprehensive 2012 study evaluated CSP using a reduced form commitment and dispatch model and quantitatively identified a number of benefits of TES (Mills and Wiser 2012). This study evaluated changes in the long-run benefits of CSP with and without TES in California using an investment model that included a "fleet-based" commitment and dispatch component for conventional generators. The study also isolates the value proposition for CSP with TES, including energy, day-ahead forecast error, ancillary service requirements, and capacity value, although the simplified commitment and dispatch component of their model did not have the fidelity to represent detailed individual unit commitment and dispatch decisions.

Several studies initiated in 2011, including the second phase of WW-SIS (Lew et al. 2012), examine CSP in greater detail. These studies use the PLEXOS production cost model and simulate the operation of the Western Interconnection in the United States. Simulation of the grid over large areas is important because of its interconnected nature and the corresponding ability of utilities to share resources over large areas. The Western Interconnection consists of thousands of generators, each of which must be simulated in detail.

Given the complexity of a large grid, it can be difficult to validate proper operation of a new generator type and isolate the cost impacts of a relatively small change in the system. As a result, to evaluate the performance of CSP, we began with a test system within a subsection of the Western Interconnection. This test system was used to evaluate the performance of CSP and the incremental value of TES under various grid conditions, including penetration of renewable generators, and compare CSP with storage to other generation sources.

9.3 IMPLEMENTATION OF CSP IN A PRODUCTION COST MODEL

9.3.1 CHARACTERISTICS OF CSP PLANTS

Two common designs of CSP plants—parabolic troughs and power towers—concentrate sunlight onto a heat transfer fluid (HTF), which is used

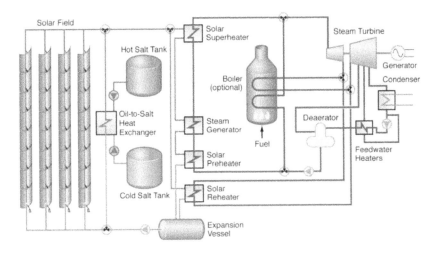

FIGURE 1: Components of a trough-type CSP plant with TES Source: EPRI (2010)

to drive a steam turbine. An advantage of CSP over non-dispatchable renewables is that it can be built with TES, which can be used to provide
multiple grid services, including shifting generation to periods with reduced solar resource.

A CSP plant with TES consists of three independent but interrelated
components that can be sized differently: the power block, the solar field,
and the thermal storage tank. Figure 1 illustrates the main components of
a trough-type CSP plant incorporating a two-tank TES system.

The size of the solar field, in conjunction with solar irradiance, determines the amount of thermal energy that will be available to the power
block. The sizing of the solar field is important because the relative size
of the solar field and power block will determine the capacity factor of
the CSP plant and the extent to which thermal energy will be utilized.
Undersizing the solar field will result in an underused power block and a
low capacity factor for the CSP plant because of the lack of thermal energy
during all hours except those with the highest solar resource. However, an
oversized solar field, when deployed without storage, can result in wasted

energy because the production from the solar field may exceed the power block capacity during many hours. The size of the solar field can either be measured in the actual area of the field or by using the concept of a solar multiple, which normalizes the size of the solar field in terms of the power block size. A solar field with a solar multiple of 1.0 is sized to provide sufficient energy to operate the power block at its rated capacity under reference conditions (in this case 950 W/m² of direct solar irradiance at solar noon on the summer solstice). The collector area of a solar field with a higher or lower solar multiple will be scaled based on the solar field with a multiple of one (i.e., a field with a solar multiple of 2.0 will cover roughly twice the collector area of a field with a solar multiple of 1.0).

The size of storage is measured by both the thermal power capacity of the heat exchangers between the storage tank and the HTF (measured in MW-t) and the total energy capacity of the storage tank. The power capacity of the thermal storage will equal some fraction of the maximum solar field output. The energy capacity of the storage tank is commonly measured in terms of hours of plant output that can be stored. Thermal storage allows an oversized solar field and a higher plant capacity factor than a plant without storage.

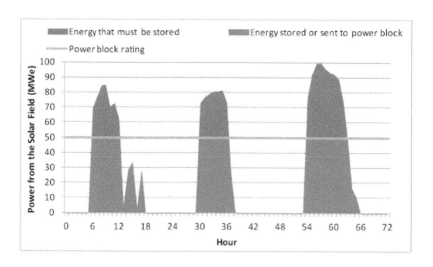

FIGURE 2: Impact of solar multiple on energy flow in a CSP plant

The concept of solar multiple, and its relation to the role of energy storage, is shown in greater detail in Figure 2, depicting the hourly flow of energy from a plant with a solar multiple of 2.0. The maximum thermal output from the solar field during any hour is 100 MWe, but the power block rating is 50 MW, meaning that energy that exceeds the power block rating must be stored for use at a later time regardless of the instantaneous demand for electricity or other grid conditions.

9.3.2 IMPLEMENTATION OF CSP IN PLEXOS

The primary function of a production cost model is to determine which generators must be committed and dispatched during each interval of the simulation and the associated cost of operation. Simulation intervals are typically 1 hour (as performed in this analysis), but there is increased interest in sub-hourly simulations, especially in scenarios of increased VG penetration where sub-hourly net load variability can require increased dispatch flexibility (CAISO 2011). A simple dispatch is determined by "stacking" generators in order of production cost (from lowest to highest) until the sum of the individual generator output is equal to load in each time interval. The actual dispatch is complicated by the many additional constraints imposed by individual generators, such as minimum up and down times and ramp rates. The actual dispatch also depends on the need for system security, including spinning reserves, which consist of partially loaded generators with the ability to rapidly ramp in response to a generator outage or unexpected increase in demand. To determine the optimal dispatch requires detailed information for each generator. Primary characteristics include maximum capacity, minimum stable output level, plant heat rate (ideally as a function of load), fuel cost, ramp rates, start time, and minimum up and down time. The software then co-optimizes the need for energy and reserves subject to the various constraints and finds the least-cost mix of generators in each time interval.

VG plants with little or no variable cost are typically placed into production cost models as a fixed hourly generation profile. Because they have no variable cost, and may also have production incentives, they are typically dispatched first but may be curtailed when operational constraints do not allow the system to accept their output. These constraints might occur when the VG exceeds the local capacity of the transmission network or during periods when conventional generators have reduced output to their minimum generation levels. This second phenomenon can be referred to as a "minimum generation" problem (Rogers et al. 2010) or an "over generation" problem (CAISO 2010). CSP plants without storage can be placed into the model in the same manner as a wind or PV plant, using the hourly output from a CSP simulation model.

For this study, trough CSP plants (both with and without storage) and PV were simulated using the System Advisor Model (SAM) (Gilman et al. 2008; Gilman and Dobos 2012) version 2012-5-11. The CSP simulations used the wet-cooled empirical trough model (Wagner and Gilman 2011). The model converts hourly irradiance and meteorological data into thermal energy and then models the flow of thermal energy through the various system components, such as losses in the HTF, finally converting the thermal energy into net electrical generation output. The CSP plant without storage assumes a solar multiple of 1.3, the SEGS VIII default power block with turbine over-design operation allowed at 105% and used default settings for all parameters, such as parasitics. Meteorological data was derived from the National Solar Radiation Database from 2006 (NREL 2007).

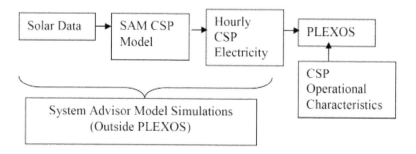

FIGURE 3: General process of implementing CSP

CSP with TES was implemented in this study using a two-step process. First, hourly electrical energy from the CSP plant was simulated using SAM, and then the electrical energy was dispatched in PLEXOS using a combination of algorithms that largely existed within that model. This process is illustrated in Figure 3 and described in more detail in the following sections.

The first step (hourly electrical energy) was created using SAM in a manner similar to the case without storage with several important differences. Essentially all parameters that are affected by plant dispatch were moved out of SAM and into the PLEXOS framework. First, the solar multiple was set to 1; a larger solar multiple and storage was implemented in the PLEXOS model as described later in this section. Second, the minimum generation levels and start-up energy requirements were set to 0 and also accounted for in PLEXOS. Parasitics were removed from the gross CSP generation to derive a net hourly generation. Operational parasitics calculated by SAM were subtracted from the electrical profile in a manner similar to other thermal power stations. We also considered the constant parasitic loads (e.g., associated with fluid pumps) that occur even when the plant is not operating. This means that the CSP plant will draw a small amount of energy from the grid and incur a small associated cost. This constant load was calculated separately based on SAM CSP simulations.

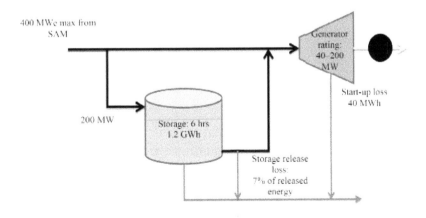

FIGURE 4: The flow of energy through a trough CSP plant with TES in PLEXOS

The product of the SAM simulation is "raw" electrical energy output, which is then processed in PLEXOS using a modified form of the PLEXOS hydro algorithm to simulate storage, generator operation, and the effect of an oversized solar field. In each hour, the model can send the electrical energy from the SAM simulations directly to the grid via a simulated power block, to storage, or a combination of both. The model can also choose to draw energy from storage. The simulated power block includes the essential parameters of the CSP power block, including start-up energy, minimum generation level, and ramp-rate constraints. The model considers start-up losses in the dispatch decision by assuming a certain amount of energy (equivalent to 20 MWhe for a 100-MW power block) is lost in the start-up process.

In CSP plants that use indirect storage, the additional efficiency losses in the storage process are also simulated. The storage losses are set to 7%, which capture both the efficiency losses in the heat exchangers and the longer-term decay losses. Constant parasitics were added by placing a constant load on the same bus as the CSP plant. The general implementation is illustrated in Figure 4, representing a 200-MW CSP plant with a solar multiple of 2.0 and 6 hours of storage.

Figure 4 also shows the effect of solar multiple, which is captured in the sizing of the power block and storage components. For example, a solar multiple of 2.0 can be simulated by setting the maximum size of the power block to 50% of the maximum output from the CSP simulations from SAM. Likewise, the storage system can be sized to accommodate some fraction of the maximum CSP output. The storage energy capacity (hours of storage) can be set independently.

9.4 TEST SYSTEM

CSP was implemented in a test system to better verify the basic performance of this dispatchable energy source and to more easily isolate the relative value of TES under various scenarios.

The best locations for CSP in the United States are in the desert southwest within the Western Interconnection. Simulating the entire interconnection makes it difficult to isolate the performance of CSP, so a smaller test system was created to develop and validate the modeling approach.

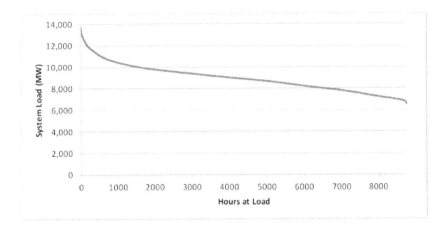

FIGURE 5: Load duration curve in 2020 for the PSCO/WACM test system

Most of the existing and proposed CSP is in California; however, simply running the California system in isolation ignores the substantial interconnections between the California and bordering states. As an alternative, we developed a system composed of two balancing areas largely in the State of Colorado: Public Service of Colorado (PSCO) and Western Area Colorado Missouri (WACM). These balancing areas consist of multiple individual utilities and this combined area is relatively isolated from the rest of the Western Interconnection. In addition, Colorado has sufficient solar resource for CSP deployment in the San Luis Valley in the south-central part of the state, and there have been proposals for large-scale solar development in the area (Xcel 2011). The test system also has sufficient wind resources for large-scale deployment, which makes evaluation of high renewable scenarios more realistic.

The Colorado test system was isolated by physically "turning off" the generation and load and aggregating the transmission outside of the PSCO and WACM balancing areas. Transmission was modeled zonally, without transmission limits within each balancing authority area. It is very difficult to simulate any individual or group of balancing authority areas as actually operated because the modeled system is comprised of vertically integrated utilities that balance their system with their own generation and bilateral transactions with their neighbors that are confidential. Not having access to that information, we modeled the test system assuming least-cost economic dispatch. We based our inputs and assumptions as much as possible on the Western Electricity Coordinating Council (WECC) Transmission Expansion Policy Planning Committee (TEPPC) model and other publicly available datasets. Projected generation and loads were derived from the TEPPC 2020 scenario (TEPPC 2011). Hourly load profiles were based on 2006 data and scaled to match the projected TEPPC 2020 annual load. The system is a strongly summer peaking system with a 2020 coincident peak demand of 13.7 GW and annual demand of 79.0 TWh. The system load duration curve for 2020 is shown in Figure 5.

The generation dataset was derived from the TEPPC 2020 database and included plant capacities, heat rates, outage rates (planned and forced), and several operational parameters, such as ramp rates. A total of 201 thermal and hydro generators were included in the test system, with total capacities listed in Table 2. The generator database was modified to include part-load heat rates based on Brinkman et al. (2012). Start-up costs were added using the start-up fuel requirements in the generator database plus the operations and maintenance (O&M) related costs based on estimates prepared for the WWSIS II study (Intertek/APTECH 2012). We adjusted the generator mix to achieve a generator planning reserve margin of 15% by adding a total of 1,450 MW (690 MW of combustion turbines and 760 MW of combined cycle units).

TABLE 1: Characteristics of the Test System Conventional Generators in 2020

	System Capacity (MW)
Coal	6,178
Combined Cycle (CC)	3,724
Gas Turbine/Gas Steam	4,045
Hydro	773
Pumped Storage	560
Other[a]	513
Total	15,793

[a]Includes oil and gas-fired internal combustion generators and demand response.

Two renewable energy scenarios were created by adding wind and solar generation. PV profiles were generated using the SAM model with 2006 meteorology. Wind data was derived from the WWSIS dataset. A low renewable energy (RE) case was created by adding wind and solar to achieve a penetration of about 13% on an energy basis. This is a relatively small increase over the renewable penetration in 2011; Colorado received about 12% of its electricity from wind in the year ending June 2012 (EIA 2012). We also considered a high RE case where wind and solar provide about 35% of the region's energy. In each case, discrete wind and solar plants were added from the WWSIS data sets until the installed capacity produced the targeted energy penetration. The sites were chosen largely based on capacity factor, and do not necessarily reflect existing or planned locations for wind and solar plants. Table 2 lists characteristics of the system in the two cases, while Table 3 provides additional details of the renewable and conventional generation mix.

TABLE 2: Renewable Scenarios in the Test System in 2020

	System Capacity (MW)	
	Low RE Scenario	High RE Scenario[a]
Wind Capacity (MW)	3,054	6,489
Wind Energy (GWh)	9,791	20,210
Solar Capacity (MW)	395	3,630
Solar Energy (GWh)	625	6,493

[a] This is the potential generation and does not include curtailment that results in actual dispatch. About 31 GWh of wind and 23 GWh of solar were curtailed in the base high RE scenario.

Three classes of ancillary service requirements were included. The contingency reserve is 810 MW based on the single largest unit (Comanche 3). This reserve is allocated with 451 MW to PSCO and 359 MW to WACM, with 50% met by spinning units. Regulation and flexibility reserve requirements were calculated based on the statistical variability of net load described by Ibanez et al. (2012). Reserves were modeled as "soft constraints," meaning the system was allowed to not meet reserves if the cost exceeded $4,000/MWh. This high cost could result during periods where a power plant would need to start up for a very short period of time just to provide reserves. Load was also modeled as a soft constraint, with a loss-of-load cost of $6,000/MWh (though the reserve margin was adequate to avoid lost load).

Fuel prices were derived from the TEPPC 2020 database. Coal prices were $1.42/MMBTU for all plants. Natural gas prices varied by plant, and for most plants were in the range of $3.9/MMBTU to $4.2/MMBTU, with a generation weighted average of $4.1/MMBTU. This is slightly lower than the EIA's 2012 Annual Energy Outlook projection for the delivered price of natural gas to the electric power sector in the Rocky Mountain region of $4.46/MMBTU in 2020 (EIA 2012). Sensitivity to natural gas price was also analyzed.

TABLE 3: Base Case Results

	Low RE	High RE
Total Production Cost (M$)	1,491.37	1,024.38
Average Production Cost ($/MWh)	18.9	13.0
Total Generation (GWH)[a]	78,957	79,098
Generation Mix		
Coal	58.8%	52.0%
Gas Combined Cycle (CC)	20.7%	7.2%
Gas Combustion Turbine		
(CT)/Gas Steam	1.4%	1.1%
Hydro	4.8%	4.8%
Wind	12.4%	25.5%
Solar PV	0.8%	8.2%
Other	1.1%	1.2%
Fuel Use (1,000 MMBTU)		
Coal	490,923	434,426
Gas	140,447	53,928

[a]While the load is the same, the total generation is slightly different (by about 0.2%), due primarily to different operation of the pumped hydro units.

Both cases were run for 1 full year (2020, with 2006 meteorology and load pattern). The model run begins with two scheduling models to determine outage scheduling and allocate certain limited energy resources. The model then performs a chronological hourly security-constrained unit commitment and economic dispatch to minimize the overall production cost under operational and system constraints. The model performs a 24-hour ahead commitment with an additional 24-hour look-ahead period, allowing the model to effectively optimize storage utilization over a 48-hour period. The analysis in this report was performed using PLEXOS version 6.207 R01, using the Xpress-MP 23.01.05 solver, with the model performance relative gap set to 0.5%.

Table 3 provides a summary of the operational results for the two base simulations. This represents on the variable cost of system operation, dominated by the cost fuel for thermal power plants. There was no loss of load and a small number of reserve violations (less than 40 hours per year in both cases).

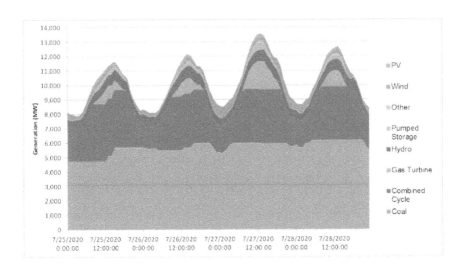

FIGURE 6. Dispatch stack during the period of July 25–28 in the low RE Case

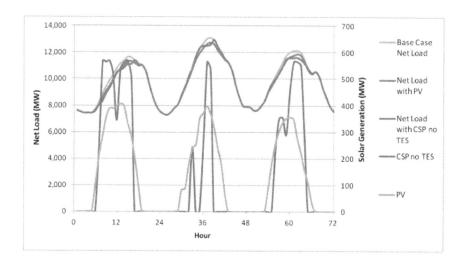

FIGURE 7: Dispatch stack during the period of February 8–11 in the high RE Case

System dispatch stacks can provide additional insight into system operation. The generation mix and dispatch was as expected, with coal units operating as baseload units, and CC and CT units operating as mid-merit and peaking units as needed. Figure 6 shows the dispatch stack for the low RE case during the week of peak demand in the summer. This figure demonstrates the opportunity for mid-day solar generation to reduce the use of the highest cost generators.

The high RE case removes much of the gas generation from the system and leaves coal on the margin for a large number of hours. Figure 7 shows a four-day period in February, which includes the day of the lowest net demand on the system in the high RE scenario. In the first two days, coal generators reduce their output to minimum levels and renewable generation is curtailed. Any additional renewable generation from 1 p.m. to 3 p.m. on February 8 and from 11 a.m. to 2 p.m. on February 9 will be unusable in this scenario, and likely curtailed. Generation during many other periods will offset mainly lower-cost coal generation. Also of note is the rapid increase in net demand that occurs after 3 p.m. when the decrease

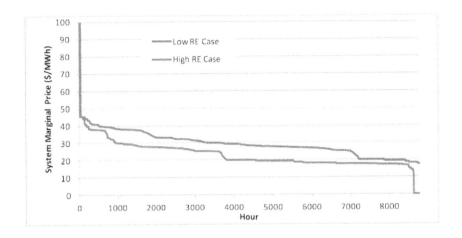

FIGURE 8: System marginal price duration curve in the PSCO balancing area for the two cases

in solar output and increase in electricity demand require large ramps of the coal units, use of higher-cost combustion turbines, and dispatch of the pumped storage plants. Previous integration studies such as WWSIS have found significant increases in ramping requirements of coal units, and a major focus of the second phase of WWSIS is to examine the potential cost implications of increased unit cycling.

In any hour of the year, the value of solar or other incremental generation in this system is determined by the marginal generators and associated price. Figure 8 is a price duration curve for the test system showing system marginal cost for the PSCO balancing area. The marginal prices for the WACM balancing area were almost identical because transmission constraints between the PSCO and WACM system were not binding; only very small price differences occurred in a few hours due to different reserve requirements. The price duration curve shows three main "zones" of prices, based on the marginal generators: coal at about $17–$20/MWh, combined cycle units at about $25–$35/MWh, and combustion turbines at about $38–$45/MWh. In the low RE case, coal is on the margin about for about 1,600 hours, while in the high RE case, coal is on the margin for over 5,000 hours. In addition, renewable energy is effectively on the margin for about 100 hours of the year in the high RE case; additional renewable generation during these hours would likely be curtailed and provides no incremental benefit to the system. There are also a small number (less than 40 hours per year) of extremely high prices, set by the reserve violation conditions.

9.5 VALUE OF CSP

The value of CSP was determined by adding a relatively small amount of generation to the base system and evaluating the change in production costs and the value of system capacity. We also examined the operation of CSP with TES, including how energy was shifted over time. Several technologies were evaluated; for uniform comparison, each technology provided an approximately equivalent amount of energy, based on the annual production of a 300-MW CSP/TES plant with a solar multiple of 2.0. This is about 1,070 GWh, or enough to provide about 1.4% of the test system demand.

9.5.1 SOLAR SCENARIOS EVALUATED

Four main cases were added to the test system, each providing an approximately equal amount of energy.

1. A flat block of zero-cost energy. This provides a point of comparison to examine how the temporal characteristics of solar energy compare to a constant or "baseload" resource. To provide an equal amount of energy, the capacity of this block was set to 123 MW.
2. Solar PV. Discrete-sized 1-axis tracking PV plants generated for the WWSIS project were added in each case until the generation equaled about 1,070 GWh, with a total installed capacity of about 580 MW. The total operating reserve requirement was also increased due to the additional short-term variability.

3. CSP no storage. A single 568-MW plant with a solar multiple of 1.3 was added to the San Luis Valley in southern Colorado. This plant produced 1,130 GWh.
4. CSP with 6 hours of storage. This case adds a 300-MW CSP plant with a solar multiple of 2.0, as discussed previously. The actual amount of energy delivered to the system varied slightly in each scenario based on the amount of energy stored (due to storage losses).

Each of the cases was simulated in both the low and high RE scenarios. Several sensitivity scenarios were considered, as discussed in Section 5.4.

9.5.2 OPERATIONAL VALUE

The operational value of each technology represents its ability to avoid the variable cost of operation. These costs were tracked in three cost categories—operating fuel, variable O&M, and start-up costs. Operating fuel includes all fuel used to operate the power plant fleet while generating and includes the impact of variable heat rates and operating plants at part load to provide ancillary services. Start-up costs include both the start fuel, as well as additional O&M required during the plant start process. In each case the operational value was calculated by dividing the total avoided generation cost in each cost category by the total potential solar generation. Table 4 summarizes the results from the production simulations.

TABLE 4: Operational Value of Simulated Generators

| | Marginal Value ($/MWh) | | | | | | | |
| | Low RE | | | | High RE | | | |
	Flat Block	PV	CSP (no TES)	CSP (6-hr TES)	Flat Block	PV	CSP (no TES)	CSP (6-hr TES)
Fuel	31.7	35.2	33.9	37.7	22.6	21.2	18.7	31.1
Variable O&M	1.2	1.0	1.0	0.8	2.1	2.0	1.9	1.4
Start	0.4	0.4	0.6	3.5	0.5	-0.9	-1.7	3.1
Total	33.3	36.6	35.5	42.1	25.2	22.3	18.9	35.6

Table 4 demonstrates three significant findings: (1) at low penetration, the value of solar generation technologies is greater than the constant (flat block) resource; (2) the value of all generation decreases as a function of renewable penetration, but the value of non-dispatchable solar resources decreases at a greater rate than the flat block or dispatchable CSP; (3) the value of CSP with storage is higher than solar technologies without storage. The range of values for different generation technologies largely can be explained by understanding the avoided fuel mix in the two different scenarios.

In the test system, the added generators (flat block or solar) reduce the output from a mix of generator types and with different efficiencies, depending on the time of day and season. Figures 9–16 illustrate how the relative value of a renewable generator is affected by the varying marginal generators and the dispatchability of the resource.

Figure 9 illustrates the relationship between price and net load for a 3-day period starting on January 22. The net load is the normal load minus wind and solar PV generation and reflects the load that must be met by other (mostly fossil fueled) generators with non-zero generation cost. The figure illustrates three zones of prices, which are seen earlier in the price

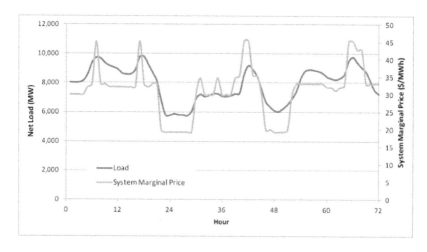

FIGURE 9: System net load and marginal price for January 22–24 (low RE case)

duration curves in Figure 8. The lowest price occurs in the overnight periods at the beginning of days 2 and 3 when coal is the marginal generator with total incremental cost of about $20/MWh. During much of the middle of the day, combined cycle units are the marginal generators, with variable costs of about $30–$35/MWh. In several periods in the morning and evening, there is an increase in net demand, where the high ramp rate or the relatively short period of increased demand requires the use of combustion turbines, resulting in a price spike to about $45/MWh. Any renewable generator added to this mix will offset energy within these three price zones but with a value depending on the temporal pattern of its output.

Figure 10 keeps the marginal price curve but adds the generation profile for CSP with and without storage. The CSP dispatch is isolated from cases where CSP is added. The total generation by these two plants is very similar, but CSP with storage is dispatched during the highest cost periods. In much of the winter, the price of electricity peaks in periods where solar output is low or zero (the morning and evening). This corresponds to when higher-cost gas-fired units are started and ramped to meet peak demand. PV and CSP without storage are unable to generate during this period and typically offset more efficient gas-fired units.

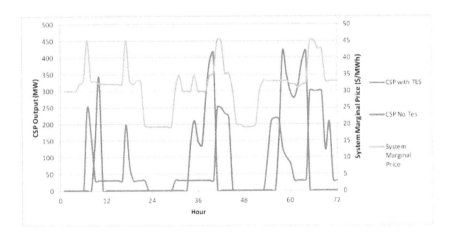

FIGURE 10: System marginal price and corresponding CSP generation on January 22–24 (low RE case)

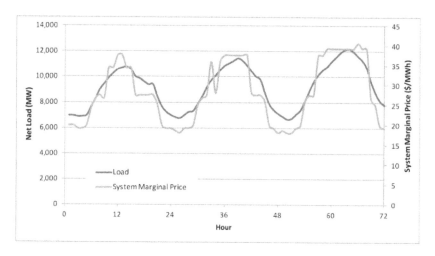

FIGURE 11. System net load and marginal price for July 14–16 (low RE case)

Alternatively, CSP with TES is able to shift generation to the evening and carry over energy to start and pick up the morning load ramp that occurs before significant solar energy is available. As a result, CSP avoids the use of higher-cost and lower-efficiency gas-fired units, producing overall higher value to the system.

During the summer, operation of CSP with storage is more continuous due to higher solar output and a different load and price profile. Figure 11 shows the relationship between net load and system marginal price for a 3-day period starting on July 14.

The corresponding CSP operation is shown in Figure 12. There are several operational issues that affect the overall and relative value of CSP with TES. First, CSP with TES is able to operate more continuously and avoid the impact of cloud cover that reduces output and increases the variability of the plant without TES. Second, CSP is able to start earlier in the day and help pick up the early morning load ramp. Finally, CSP is able to continue operation longer into the late afternoon and early evening. This is particularly important for the plant capacity value discussed in Section 5.3.

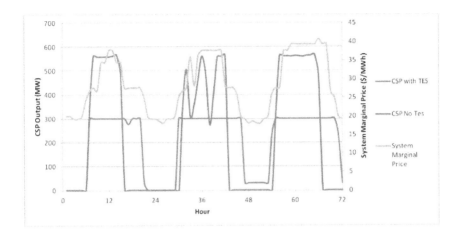

FIGURE 12: System marginal price and corresponding CSP generation on July 14–16 (low RE case)

Figure 12 shows the impact of the solar multiple, which can provide some disadvantages at low solar penetration. At low penetration of solar, PV and CSP without storage are largely coincident with demand (and relatively high prices) during the summer. As illustrated in Figure 2, whenever the thermal output of the solar field exceeds the power block capacity, energy must be stored, regardless of the system demand for energy or price. As a result, the plant is forced to store this energy and generate at a later time, even if this later time has a lower demand or lower cost of energy. This is shown in Figure 12 on the first and second day, when during some hours, CSP without storage sells more energy at periods of high prices than CSP with TES. CSP with storage is forced to shift some energy to the evening when prices are slightly lower.

The value of solar and dispatchable CSP is strongly dependent on the mix of generator types and amount of renewable energy. As the penetration of renewables increases, the patterns of net demand for electricity change, and different mixes of generation are needed to address the increasing variability and uncertainty of the wind and solar supply. Figure

13 is a duplication of Figure 9, showing price and load during 3 days in January, except for the high RE case. The large amounts of wind and solar PV have suppressed the marginal price, and coal is on the margin for more hours. The load shape (and price) is also much more volatile, with operation of combustion turbines to address the shorter peaks.

In the high RE scenario the absolute value of all energy sources drops due to lower system marginal prices. However, the value of variable energy sources drops at a much faster rate than dispatchable sources, as a plant with TES is able to change output to capture the remaining periods of high prices. Figure 14 shows how CSP with TES is able to generate during the hours of highest price during this period in January.

Dispatchability becomes increasingly important during periods of very high renewable output to avoid generating during periods of zero value and associated renewable curtailment. Figure 15 shows a period of low net demand due to high solar (and wind) output during the middle of the day. During the first 2 days shown shortly after noon, the net load drops to the point where all thermal generators have reduced output to their minimum.

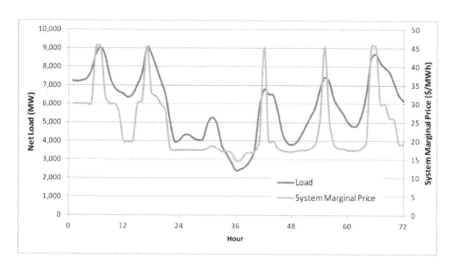

FIGURE 13: System net load and marginal price for January 22–24 (high RE case)

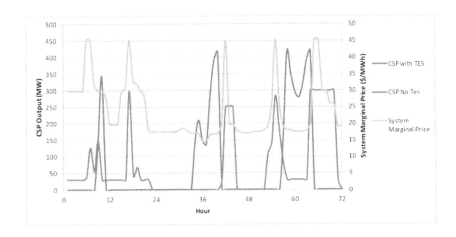

FIGURE 14: System marginal price and corresponding CSP generation on January 22–24 (high RE case)

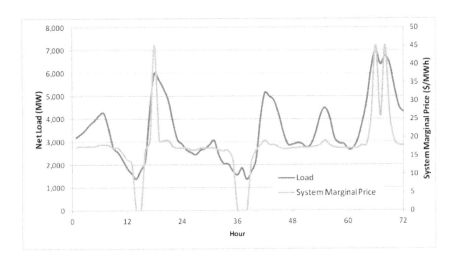

FIGURE 15: Net load and price for a 3-day period starting February 8

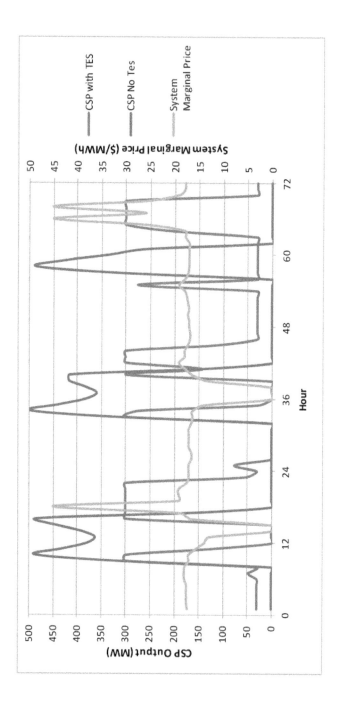

FIGURE 16: System marginal price and corresponding CSP generation on February 8–10 (high RE case)

This is the same period as the first 3 days in Figure 8, where during the middle of the day all coal plants in the system cannot reduce output further without incurring a costly shut down. Any additional zero-cost renewable energy generated during these hours cannot be used by the system so have zero value, and the system marginal price is $0/MWh.

Figure 16 shows the operation of CSP plants in these three days. CSP without storage generates in the middle of the day, producing some output that provides zero incremental value (when the system marginal price is zero). During these periods, CSP with thermal storage generates at low output, or shuts down, avoiding curtailed energy and maximizing value by shifting energy to periods of higher net demand and providing potentially valuable ramping services.

The ability to avoid renewable generation during periods of low or zero value will be an increasingly important source of value as renewable penetration increases. In the high RE scenario, about 5% of the additional PV and about 6% of the CSP without storage has zero value and is effectively curtailed. The number of hours of zero value generation (resulting in renewable curtailment) is highly non-linear as a function of renewable penetration and would be expected to increase without additional measures to increase system flexibility (Denholm and Margolis 2007).

The sum of these factors, including the mix of generation, fuel cost, and curtailment can be translated into the source of avoided fuel costs in Table 4. Tables 5 and 6 further explain the source of avoided costs for the different generator types. Table 5 indicates the type of generation avoided by each unit of generation. In the low RE case, each kilowatt-hour of CSP without storage avoids 0.9 kWh of combined cycle generation and 0.1 kWh of combustion turbine generation. In some cases, the smoothing of the load can actually increase the use of some lower-cost generator types; in the low RE case, CSP with storage and PV can improve the system dispatch and increase low-cost coal generation slightly. The flat block results show a greater displacement of coal because it generates at constant output, including at night when coal is often on the margin. Table 5 demonstrates how, in the high RE scenario, much of the gas generation has been removed by the system, and coal is on the margin for more hours. Both PV and CSP without storage remove similar amounts of combined cycle and coal; however, CSP with storage continues to avoid mostly gas generation due to the dispatchability of the resource.

In some cases, each unit of generation removes more or less than 1 unit of thermal generation. This is due to two factors: pumped storage operation and curtailment. The flat block in the low RE case frees up coal generation to displace more costly gas plant operation via the use of pumped storage. However, because storage incurs losses, this results in a small increase in thermal generation. The opposite occurs in the low RE PV case and the CSP with storage cases. The displacement of higher-cost generation in these cases reduces the economic operation of pumped storage, decreasing storage losses and resulting in more than 1 unit of avoided generation per unit of solar generation. At higher RE penetration, solar without storage displaces less than 1 unit of thermal generation due to curtailment of renewable generators.

TABLE 5: Avoided Thermal Generation

| | Avoided Thermal Generation (kWh/kWh) | | | | | | | |
| | Low RE Scenario | | | | High RE Scenario | | | |
	Flat Block	PV	CSP (no TES)	CSP (6-hr TES)	Flat Block	PV	CSP (no TES)	CSP (6-hr TES)
Coal	0.09	-0.06	-0.03	-0.08	0.55	0.50	0.52	0.17
Gas Combined Cycle	0.78	0.99	0.91	0.79	0.39	0.50	0.39	0.72
Gas Turbine/ Steam	0.10	0.09	0.11	0.27	0.05	-0.02	-0.01	0.11
Total	0.98	1.03	1.00	1.02	0.99	0.97	0.90	1.04

While Table 5 is a useful illustration of the type of generation avoided, the ultimate cost driver is the type and amount of fuel actually displaced. Table 6 provides the actual avoided operational fuel in each scenario (in MMBTU per MWh of solar generation). Of note is the fact that the avoided fuel rate increases in the high RE scenario. This is due to the displacement of lower cost, higher heat rate coal units compared to more efficient, higher-cost gas generators. The product of the avoided fuel in Table 6 and fuel costs produce the fuel value ($/MWh) in Table 4.

TABLE 6: Avoided Fuel

| | Avoided Fuel (MMBTU/MWh) | | | | | | | |
| | Low RE Scenario | | | | High RE Scenario | | | |
	Flat Block	PV	CSP (no TES)	CSP (6-hr TES)	Flat Block	PV	CSP (no TES)	CSP (6-hr TES)
Coal	1.1	-0.7	-0.7	-0.9	5.8	5.2	5.4	1.9
Gas	7.4	8.9	8.9	9.7	3.5	3.6	2.9	7.1
Total	8.5	8.2	8.2	8.8	9.3	8.8	8.3	9.0

An additional important secondary source of value for CSP with TES is the ability to avoid thermal plant starts and associated fuel use and maintenance. Even at low penetration, PV and CSP without storage tends to increase the variability of the net load, increasing the number of plant starts but decreasing the total amount of energy produced by the generation fleet. Table 7 provides the estimated number of avoided starts and percentage reduction. Consistent with the previous tables, a positive number represents actual avoided starts (a net benefit), while a negative number means an increase in starts. This table demonstrates a significant reduction in starts due to the flexible operation of CSP with TES.

TABLE 7: Avoided Starts

| | Avoided Starts (Total/%) | | | | | | | |
| | Low RE Scenario | | | | High RE Scenario | | | |
	Flat Block	PV	CSP (no TES)	CSP (6-hr TES)	Flat Block	PV	CSP (no TES)	CSP (6-hr TES)
Coal	3/ 0.4%	-1/ -1.1%	-5/-0.7%	-8/-1.1%	-16/ -2.2%	4/ 0.6%	-3/ -0.4%	-18/ -2.5%
Combined Cycle	-77/ -6.3%	18/ 1.5%	53/4.3%	56/4.6%	9/1.2%	17/ 2.2%	-56/ -7.2%	129/ 16.5%
Gas Turbine/ Steam	362/ 4.6%	-412/ -5.2%	-271 /-3.4%	1,099/ 13.8%	432/ 4.1%	-640/ -6.1%	-361/ -3.4%	871/ 8.3%

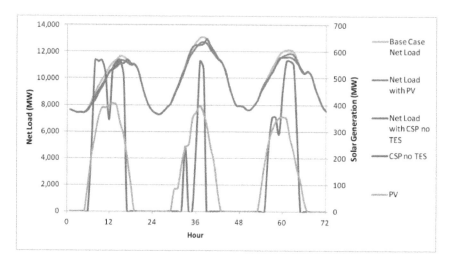

FIGURE 17: Correlation of demand and solar generation on a 3-day period starting July 26 (low RE case)

9.5.3 CAPACITY VALUE

The value calculated in Section 5.2 only addresses the variable operational value. Both CSP and PV have the ability to provide system capacity and replace new generation. However, the actual capacity value of solar technologies depends on their coincidence with demand patterns and how this coincidence changes as a function of penetration.

At low penetration, the capacity credit (equal to the fraction of capacity that is available during periods of high net demand) of PV and CSP without TES is relatively high. Figure 17 shows the simulated solar output during three peak demand days in the low RE system, including the system annual peak on July 27 showing high correlation. As a result, each megawatt of PV or CSP without TES reduces the net demand by a significant amount and eliminates the need for conventional generation.

As the penetration of PV or CSP without storage increases, the capacity credit drops significantly. Solar energy shifts the peak to later in the day, to periods where solar output is low or zero. In Figure 17, the annual peak demand occurred in the hour ending at 2 p.m. on July 27.

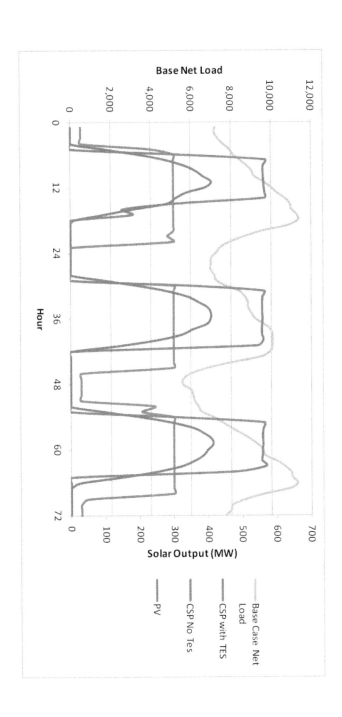

FIGURE 18: Correlation of demand and solar generation on a 3-day period starting July 17 (high RE case)

(Other peak demand hours are typically an hour or two later.) However, in the high RE case, where solar provides 8% of total demand, the net demand has been shifted to later in the day where solar is no longer highly correlated with load. Figure 18 shows an example of a new period of high peak demand in the high RE case where the net load peaks in the hour ending at 7 p.m. on the first and third day. On these 3 days beginning on July 17, there is still strong solar output, but PV and CSP without storage no longer provide significant amounts of net demand reduction. CSP with storage shifts generation to later in the day and provides a net demand reduction equal to the plant's rated capacity, resulting in a capacity credit of close to 100%. This is shown in detail in Figure 19, which enlarges the net load on July 17 and shows the net demand after removing the generation from the three different solar technologies.

Estimation of the monetary value of system capacity begins with an estimate of each plant's capacity credit. There are a number of methods used to estimate the capacity credit of VG sources. We used the simple capacity factor approximation technique, which has been shown to be a reasonable approximation for more computationally complex methods (Madaeni et al. 2012).

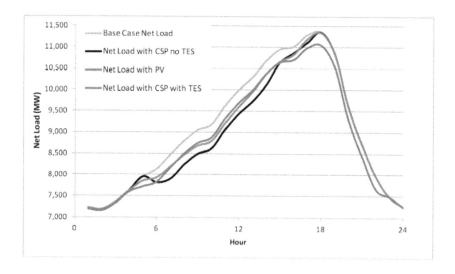

FIGURE 19: Net demand with different solar generation types on July 17 (high RE case)

Based on 2006 load and solar patterns, in the low RE case, where PV provides about 1% of total demand, each 100 MW (AC rating) of PV or CSP without storage provides about 70–75 MW of system capacity value before taking into account forced outages. This is comparable to a previous estimate of PV in Colorado (Xcel 2009). Adding TES to CSP increases the capacity value substantially.

Table 8 summarizes the capacity value estimates from this analysis. The first row in Table 8 is the capacity credit in terms of fraction of rated capacity. This value assumes an equal outage rate for maintenance across technologies. The second row translates this into an annualized value per installed kilowatt of the corresponding technology by multiplying the capacity credit by the low and high estimated annual value of a reference generator with 100% availability. The low value of the reference generator is $77/kW, based on the estimated annualized cost of a combustion turbine, while the high value is $147/kW, based on the annualized cost of combined cycle generator.

Row 3 of Table 8 translates this value per installed kilowatt into a value per unit of generation. This is calculated by multiplying the value per unit of capacity by the total capacity (to get the total annual value of the installed generator), then dividing this value by the total energy production. This introduces some unusual and somewhat counterintuitive outcomes, resulting largely from the impact of solar multiple and the use of TES, as demonstrated previously by Mills and Wiser (2012). A CSP plant with storage and a PV plant providing equal amounts of energy on an annual basis will have a different installed capacity. In the test system, 300 MW of CSP with a solar multiple of 2.0 and 6 hours of storage provides the same amount of energy as 577 MW of PV capacity. In the low penetration case, CSP provides 294 MW of system capacity at a capacity credit of 98%, while PV at a 70% capacity value provides 404 MW. This means the aggregated PV plant has a higher overall capacity value than the CSP plant, and because both plants produce the same amount of energy, PV produces a higher value of capacity on a per unit of energy basis. This effect disappears in the high RE case where the capacity value of PV and CSP without storage is very low. This issue is illustrated conceptually in Figure 20, where the output of PV and CSP is shown for a single day (July 17). It shows that at the peak hour, a CSP plant without storage has a higher capacity value than the CSP plant with storage. It also shows that this benefit on this particular day is at the very edge of production and again demonstrates the dramatic drop in capacity value of PV and CSP without storage at fairly low penetration.

TABLE 8: Capacity Value

	Low RE Scenario				High RE Scenario			
	Flat Block	PV	CSP (no TES)	CSP (6-hr TES)	Flat Block	PV	CSP (no TES)	CSP (6-hr TES)
Capacity Credit (%)	100	70	75	98	100	13	3	78
Capacity Value (Low/High) ($/kW)	77/147	54/103	58/110	76/144	77/147	10/19	3/5	60/115
Capacity Value (Low/High) ($/MWh)	8.8/16.8	29.7/56.6	29.1/55.3	21.2/40.4	8.8/16.8	5.3/10.1	1.3/2.4	17.1/32.6

In the high RE scenario, CSP with storage is able to generate at nearly full output during remaining high demand periods in the summer. However, it experiences a reduction in overall capacity value due primarily to limited energy availability during a few hours of relatively high demand in the winter.

9.5.4 TOTAL VALUE

The total value of the different generation sources is the sum of the operational value and capacity value. Figure 21 summarizes the values for the different cases by combining the operational value from Table 4 and the capacity value from Table 8.

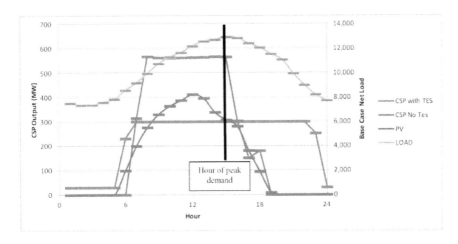

FIGURE 20: Comparison of solar output on a high-demand day (July 17)

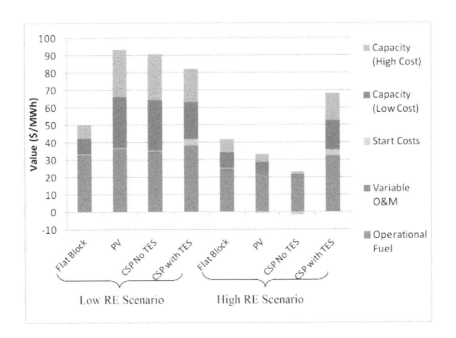

FIGURE 21: Total value of generation sources in the test system

9.5.5 SENSITIVITIES AND COMPARISON TO PREVIOUS WORK

9.5.5.1 HOURS OF STORAGE

Additional cases were run with CSP with 8 hours of storage and no changes to the solar multiple. The value of CSP increased slightly, by about $0.9/MWh in the low RE case and about $1.0/MWh in the high RE case. This decrease in marginal value of additional storage for a constant solar multiple or lower has been observed previously (Sioshansi and Denholm 2010.) However, additional analysis is needed to quantify the impact of various amounts of storage and different technology types, particularly under various RE penetration scenarios.

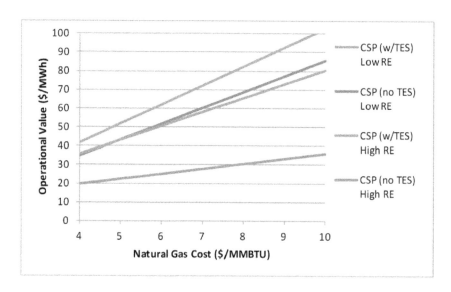

FIGURE 22: Estimated operational value of CSP as a function of gas prices

9.5.5.2 NATURAL GAS PRICE

The natural gas price used in the test system ($4.1/MMBTU) is much lower than historical prices. Given the uncertainty over fuel prices and the high level of sensitivity of the results to natural gas prices, we estimated the impact of varying the cost of natural gas. We ran scenarios with generation-weighted natural gas prices equal to $5.9/MMBTU and $7.8/MMBTU. We found that total fuel use is similar in the various cases and the value of avoided fuel is proportional to fuel price. (This relationship would not necessarily hold for lower gas prices, where at some point the dispatch stack of gas and coal generation would be inverted, and the variable cost of coal generation would put a floor on the value of avoided generation.) Additional analysis and discussion of this is provided by Diakov et al. (2012). Figure 22 provides an estimate of the change in the operational value of CSP as a function of natural gas prices for the two RE penetration cases, using a linear fit to the results from the three fuel price scenarios.

9.5.5.3 COMPARISON TO PREVIOUS ESTIMATES

The analysis of CSP in the test system shows much lower operational value for CSP compared to two previous analyses. Table 9 lists the value of CSP with and without storage from two previous studies.

TABLE 9: Previous Estimates of the Operational Value of CSP With and Without TES

Study Location	CSP Value ($/MWh)		Value Considered	Source[a]
	CSP (no TES)	CSP (with 6-hour TES)		
Arizona	47.0	50.5	Energy Only	1
New Mexico	61.2	66.2	Energy Only	1
California	58.5	67.9	Energy Only	1
Texas	89.4	98.4	Energy (with scarcity)	1
California	53.8 (energy)	56.3 (energy)	Energy + Capacity (Separate)	2

[a]Sources: 1= Sioshansi and Denholm (2010); 2 = Mills and Wiser (2012)

The differences between the test system value and previous analysis result from several factors. For appropriate comparison, the type of market or value must be considered. The first three locations in Table 9 are used historic energy-only market prices. These were derived from two regulated markets (Arizona and New Mexico) and a wholesale energy-only market (the California Independent System Operator - CAISO). In these studies, CSP was dispatched against hourly system marginal prices published by the utility or system operator. For the regulated markets, these system marginal prices are calculated in a manner similar to the dispatch model used in this study. For the CAISO market, the data is the system market clearing price calculated by the system operator. The most appropriate comparison in this study is the low RE operational value in Table 4 (equal to $35.5/MWh and $42.1/MWh for plants without and with TES, respectively). These values are lower than results from the previous studies, and the differences can be largely explained by both fuel mix and fuel price. The previous study results used electricity price data from 2005. In 2005, the average delivered price of natural gas to utilities in Colorado was $7.41/MMBTU compared to $4.1/MMBTU in the 2020 test system. Adjusting the value of the test system to this higher-priced gas value using the estimates in Figure 23 produces an approximate CSP operational value of $63/MWh for CSP without storage and $76/MWh for CSP with storage. These values are closer to the previous estimates in Table 9, with the difference likely explained by the fuel mix. The largest difference is between the test system and Arizona, where coal was on the margin for more hours than in the Colorado system where gas is on the margin for almost all hours of CSP generation.

The value for Texas is derived from an energy-only market (the Electric Reliability Council of Texas – ERCOT). However, the CSP value results from the ERCOT case are still higher than can be explained by higher natural gas prices (even after adjusting for the average Texas natural gas price of $8.1/MMBTU in 2005). This difference appears to be driven by scarcity pricing in the ERCOT market. In locations such as ERCOT that do not have capacity markets, high energy prices may be needed to recover capacity costs. These high prices will occur during peak periods where capacity reserves are short. In the ERCOT case, for example, there are over 400 hours where the price of energy exceeded $150/MWh. This means that the values from the ERCOT system represent more than the pure variable cost of CSP benefits. A more appropriate comparison would be to combine the operational value of CSP in the test system (adjusted for the difference in natural gas prices) and the capacity value. However, it is not clear that the scarcity pricing that occurred in ERCOT in 2005 would be sufficient to support new capacity, and ERCOT has changed its rules since 2005 in part to address this issue. As a result, it is difficult to directly compare the Texas results with those from the test system.

The final value in Table 9 is from the Mills and Wiser (2012) analysis in California that used a reduced form dispatch model to estimate both operational and capacity value independently. The analysis used a natural gas price of $6.4/MMBTU. Applying this value to the test system produces an estimated operational value of $55/MWh and $66/MWh for CSP plants without and with TES, respectively, which are closer to the values in the California study. The California study also estimated capacity value—it found somewhat lower capacity credit for CSP plants (particularly with storage), perhaps due to the different load profiles between the two states.

More analysis is needed to understand how the dispatch decisions will affect the operation and associated planning capacity applied to CSP plants with TES.

9.5.6 STUDY LIMITATIONS AND FUTURE WORK

This analysis did not perform a complete assessment of the value of CSP with TES. In addition to its limited geographical scope, there are several limitations to the analysis that will be addressed in future studies, including WWSIS II. A primary limitation is related to sub-hourly operation and ramp rates due to both variability and uncertainty. The simulations were performed at an hourly level and did not consider additional ramping that would result at higher time resolution due to solar variability. Furthermore, while flexibility reserves were held to account for solar forecast error, these reserves were not dispatched. This would further increase generator ramp requirements. Finally, the high ramp rates that are observed in these hourly simulations impose no additional cost penalty. WWSIS II will include 5-minute dispatch and the impact of ramping costs. These simulations should aid in identifying cost impacts on conventional generators and possible mitigation through the use of CSP with TES.

This study also did not perform a detailed assessment of the ability of CSP to provide ancillary services, including regulation, flexibility, or spinning contingency reserve. In conventional power systems, these services incur a cost due to two factors. The first is the additional O&M and other operational costs associated with additional cycling and other plant operations required to provide ancillary services. The second is the "opportunity cost" associated with holding plants at part load compared to operating at full output. We did not have sufficient data on the operational cost of CSP plants in providing ancillary services and how these costs compare to other generators to evaluate them in this study. We did perform simulations in an attempt to calculate any reduction in production cost when providing ancillary services from CSP, and found the reduction in total system cost to be extremely small. However, the modeled system has very low costs in general for provision of ancillary services due to the assumed flexibility of many generator types in the database. Additional analysis will be required to further isolate benefits of CSP providing ancillary services.

Finally, this analysis considered a single CSP technology type: wet-cooled trough-type plants with a limited range of storage capacities. Further analysis is needed to evaluate the impact of multiple CSP technology types under a range of renewable penetration scenarios.

9.6 CONCLUSIONS

Implementation of CSP with TES in commercial production simulation and planning tools is an important component of valuing this technology. This study evaluated the operation of CSP with TES in two scenarios of renewable penetration in a test system based on two balancing areas in Colorado and Wyoming. Overall, we found that the simulated CSP plants were dispatched to avoid the highest-cost generation, generally shifting energy production to the morning and evening in non-summer months and shifting energy towards the end of the day in summer months. This minimized the overall system production cost by reducing use of the least-efficient gas generators or preferentially displacing combined cycle generation over coal generation. The system also dispatches CSP during the periods of highest net load, resulting in a very high capacity value.

Overall, the addition of TES to CSP increases its value; however, the difference in value between plants with and without storage is highly dependent on both the cost of natural gas and the penetration of other renewable sources, such as PV. At low penetration of renewables, the inherent coincidence of solar and price patterns means that CSP without storage (and PV) has relatively high value. Combined with a relatively low gas price of $4.1/MMBTU used in this study results in an incremental operational value of TES of about $6.6/MWh over a plant without TES (at low RE penetration). At higher RE penetration, this difference increases as the value of mid-day generation is reduced—in the high RE test system this difference in operational value grew to $16.7/MWh. In addition, the capacity value of CSP systems with TES remains high, further increasing the difference in value associated with TES.

REFERENCES

1. Brinkman, G.; Denholm, P.; Drury, E.; Ela, E.; Mai, T.; Margolis, R.; Mowers, M. (2012). Grid Modeling for the SunShot Vision Study. NREL/TP-6A20-53310. Golden, CO: National Renewable Energy Laboratory.
2. Brinkman, G.; Lew, D.; Denholm, P. (2012). "Impacts of Renewable Generation on Fossil Fuel Unit Cycling: Costs and Emissions." PR-6A20-55828. Golden, CO: National Renewable Energy Laboratory, 42 pp.
3. California Independent System Operator (CAISO). (31 August 2010). "Integration of Renewable Resources. Operational Requirements and Generation Fleet Capability at 20% RPS." Folsom, CA: CAISO.
4. CAISO. (10 May 2011). "Summary of Preliminary Results of 33% Renewable Integration Study –2010." CPUC LTPP Docket No. R.10-05-006. Folsom, CA: CAISO.
5. Denholm, P.; Hand, M.; Mai, T.; Margolis, R.; Brinkman, G.; Drury, E.; Mowers, M.; Turchi, C. (2012). The Potential Role of Concentrating Solar Power in High Renewables Scenarios in the United States. NREL/TP-6A20-56294. Golden, CO: National Renewable Energy Laboratory.
6. Denholm, P.; Margolis, R. M. (2007). "Evaluating the Limits of Solar Photovoltaics (PV) in Traditional Electric Power Systems." Energy Policy (35:5); pp. 2852–2861.
7. Denholm, P.; Mehos, M. (2011). Enabling Greater Penetration of Solar Power Via the Use of Thermal Energy Storage. TP-6A20-52978. Golden, CO: National Renewable Energy Laboratory.
8. U.S. Department of Energy (DOE). (2012). SunShot Vision Study. BK-5200-47927; DOE/GO-102012-3037. Washington DC: U.S. Department of Energy.

9. Diakov, V.; Jenkin, T.; Drury, E.; Bush, B.; Byrne, R.; Denholm, P.; Margolis, R.; Milford, J. (2012). Reducing the Uncertainty of the Cost of Electricity to Consumers and Other Risk Mitigation Benefits of Large Scale Solar and Wind Energy Penetration. Golden, CO: National Renewable Energy Laboratory.

10. U.S. Energy Information Administration (EIA). (August 2012). "Electric Power Monthly with Data for June 2012." Washington, DC: EIA.

11. EIA. (2012). "Annual Energy Outlook." Washington, DC: EIA.

12. Electric Power Research Institute, EPRI. (2010). Solar Thermocline Storage Systems: Preliminary Design Study. Electric Power Research Institute, Palo Alto, CA: June 2010. 1019581.

13. GE Energy. (2010). Western Wind and Solar Integration Study. SR-550-47434. Golden, CO: National Renewable Energy Laboratory, 536 pp.

14. Gilman, P.; Blair, N.; Mehos, M.; Christensen, C.; Janzou, S.; Cameron, C. (2008). Solar Advisor Model User Guide for Version 2.0. TP-670-43704. Golden, CO: National Renewable Energy Laboratory, 133 pp.

15. Gilman, P.; Dobos, A. (2012). System Advisor Model, SAM 2011.12.2: General Description. TP-6A20-53437. Golden, CO: National Renewable Energy Laboratory, 18 pp.

16. Ibanez, E.; Brinkman, G.; Hummon, M.; Lew, D. (2012). A Solar Reserve Methodology for Renewable Energy Integration Studies Based on Sub-Hourly Variability Analysis. Golden, CO: National Renewable Energy Laboratory.

17. Intertek APTECH. (April 2012). "Power Plant Cycling Costs." EAS 12047831-2-1. Accessed November 5, 2012: http://wind.nrel.gov/public/WWIS/APTECHfinalv2.pdf.

18. Kirby, B. (December 2004). "Frequency Regulation Basics and Trends." ORNL/TM 2004/291. Oak Ridge, TN: Oak Ridge National Laboratory.

19. Lew, D.; Brinkman, G.; Ibanez, E.; Hodge, B.M.; King, J. (2012). Western Wind and Solar Integration Study Phase 2: Preprint. CP-5500-56217. Golden, CO: National Renewable Energy Laboratory, 9 pp.

20. Madaeni, S. H.; Sioshansi, R.; Denholm, P. (2012). Comparison of Capacity Value Methods for Photovoltaics in the Western United States. TP-6A20-54704. Golden, CO: National Renewable Energy Laboratory, 38 pp.

21. Mai, T.; Wiser, R.; Sandor, D.; Brinkman, G.; Heath, G.; Denholm, P.; Hostick, D.J.; Darghouth, N.; Schlosser, A.; Strzepek, K. (2012). Renewable Electricity Futures Study. Volume 1: Exploration of High-Penetration Renewable Electricity Futures. (Volume 1 of 4). Golden, CO: National Renewable Energy Laboratory; NREL Report No. TP-6A20-52409-1.

22. Mills, A.; Wiser, R. (June 2012). Changes in the Economic Value of Variable Generation at High Penetration Levels: A Pilot Case Study of California. LBNL-5445E. Berkeley, CA: Lawrence Berkeley National Laboratory.

23. NREL. (April 2007). National Solar Ration Database 1991-2005 Update: User's Manual. NREL/TP-581-41364. Golden, CO: National Renewable Energy Laboratory.

24. Public Service Company of Colorado (PSCO). (October 2011) 2011 Electric Resource Plan. Volume II Technical Appendix.

25. Rogers, J.; Fink, S.; Porter, K. (2010). Examples of Wind Energy Curtailment Practices. SR-550-48737. Golden, CO: National Renewable Energy Laboratory.
26. Sioshansi, R.; Denholm, P. (2010). The Value of Concentrating Solar Power and Thermal Energy Storage. NREL/TP-6A2-45833. Golden, CO: National Renewable Energy Laboratory.
27. TEPPC. (September 2011). "TEPPC 2010 Study Program 10-Year Regional Transmission Plan 2020 Study Report." Salt Lake City, UT: WECC.
28. Usaola. (2012). "Participation of CSP Plants in the Reserve Markets: A New Challenge for Regulators." Energy Policy. (49); pp. 562–571.
29. Wagner, M.; Gilman, P. (June 2011). Technical Manual for the SAM Physical Trough Model. NREL/TP-5500-51825. Golden, CO: National Renewable Energy Laboratory.
30. Xcel. (February 2009) "An Effective Load Carrying Capability Analysis for Estimating the Capacity Value of Solar Generation Resources on the Public Service Company of Colorado System." Minneapolis, MN: Xcel Energy Services, Inc.
31. Xcel. (31 October 2011). "Public Service Company of Colorado 2011 Electric Resource Plan." Minneapolis, MN: Xcel Energy Services, Inc.

.

CHAPTER 10

THE PLACE OF SOLAR POWER: AN ECONOMIC ANALYSIS OF CONCENTRATED AND DISTRIBUTED SOLAR POWER

VANESSA ARELLANO BANONI, ALDO ARNONE,
MARIA FONDEUR, ANNABEL HODGE, .J PATRICK OFFNER,
AND JORDAN K. PHILLIPS

10.1 BACKGROUND

Carbon-based fuel sources are becoming a hot commodity as the domestic electric industry watches the future. Proponents of renewable energy argue that an alternative approach will allow for more sustainable energy usage, help support future growth, avoid price spikes, allow for energy independence, and ultimately help slow the progression of global warming. To help illustrate such approach, consider a solar farm composed of Stirling engines covering an area of 100 squared miles. This alone could replace all the coal burned to generate energy in the United States [1].

Despite these positive externalities, the potential of major cost inequality and the associated fixed costs of renewable resources fuels debates. Renewable energy must combat the already present, tested, cheap, and ultimately reliable methods currently used to generate power.

This paper examines the cost and benefits, both financial and environmental, of two leading forms of solar power generation, grid-tied photovoltaic cells (PVs) and Dish Stirling Systems (DSS), using conventional carbon-based fuel as a benchmark. First, it will establish the manner in which these technologies, PVs and DSS, will be implemented in our study. Secondly, it will define a model city, its location, characteristics and constraints, which will be used as a parameter to evaluate the benefits and costs of each technology. Finally, it will attempt to determine whether decentralized photovoltaic farming is more effective and sustainable than a central, Stirling-engine based solar farm for our model city, with calculations related to fixed costs (construction, core technology used, land) and variable costs (labor, upkeep) determining the final prices of each power source. Our ultimate conclusion will be based on which power source is better from a consumer standpoint.

10.2 METHODS

10.2.1 SETTING CONCENTRATED SOLAR POWER AND DISTRIBUTED SOLAR POWER EXEMPLIFIERS

As the still immature solar energy market has grown we have learned more about different technologies and their ideal application. On the one hand, the flat panel photovoltaic cells, typically made of silicon, are the best-known form of solar technology [2], while the Concentrated Solar Power (CSP) industry is still at its infancy. While both provide a means of electricity production, this study is concerned with finding out which is the optimal means of energy consumption for a standard, West Coast suburban area.

When designing the large scale, high-priced solar farms, CSP is much preferred due to its cost effectiveness. However, CSP requires a large amount of room and very large-scale equipment to be most effective. Additionally, most recent plant installations have shown that economies of scale are applicable and therefore, as plant size increases, capital costs decrease [3]. Given this information we have chosen one of the most promising technologies, the Dish Stirling system, as our large-scale electricity producer.

Comparatively, photovoltaic energy production is far more effective when used in a decentralized manner due to its intrinsic properties, like its smaller size, which allows for more flexibility in the size of an installation. The household installable PV cells allows for single home power generation, with a surplus sent back to the grid for profit. These cells, though expensive, are often accompanied by a tax incentive. This allows for an analysis of decentralized means of power production without the large scale fixed costs of a central producer.

10.2.2 CREATION OF A MODEL CITY

Our goal is to get an accurate representation of the power needs and consumer habits of a typical city. In order to better account for variances and external influences, such as city demographics and weather, we decided to create a model city to test the two methods of solar-powered electric distribution.

10.2.3 MODEL CITY LOCATION AND THE POTENTIAL OF THE SUN

The United States is of considerable interest for this study as it receives an enormous amount of solar heat when compared to the rest of the world. Each year the Earth intercepts a large amount of radiant heat, equaling roughly 5×10^{20} kilocalories. Thought of in terms of area, a typical square foot of land in the United States receives more than 1 kilocalorie per square foot, per minute or 500 kilocalories per day. Aggregated over an acre, those 40,000 square feet receive 20,000,000 kilocalories per day. Now, a conservative estimate for energy usage derived from coal, barrels of oil, and cubic feet of gas is somewhere around 150,000 kilocalories per day. When compared with the above stated estimate for light energy, the Sun could supply 2,000 times the heat energy currently used in the United States. Though promising, the illusive issue still remains, turning the potential energy into useful, usable electric energy.

It becomes very obvious that location is of prime importance for successful solar farming and energy production. The intensity of solar radiation outside of the Earth's atmosphere is about 1,300 watts per square meter. We must assume that some of this is lost in the haze and cloud cover, leading to an estimate of 80-90% of the solar radiation successfully entering the atmosphere and reaching the ground. For simplicity we estimate this amount to be 1.100 kilowatts per square meter. The composition of light that enters is also of great importance, as it determines the applicable technology. The rays of sunlight are composed of diffuse light (scattered) and also direct rays from the sun (normal radiation). The above factors of haze, humidity and cloud cover can affect the light distribution and lead to increased scattering. As described by Leitner, flat panel PV power plants use both diffuse and direct radiation, while CSP can only harness the direct sunlight.

California is a prime location due to its latitude, low cloud cover and humidity, and the amount of sunlight received, as well as its great government incentives. This also works for PV, but further modifiers are required for the proposed solar farm. The large land requirements are not difficult to find, especially in the Western deserts of the United States. Not only is space plentiful, but also the conditions are ideal. This land required must be flat, as well as corresponding with other potential limiting factors. These factors, which affect the size of the land available, include military bases, national parks and protected wilderness, cropland, and developing urbanization. According to Leitner, land can be categorized into three resource classes of average solar energy resource (kWh/m²/day): 6.0 to 6.5 (good), 6.5 to 7.0 (great), and 7.0 and above (excellent). Given these factors, careful analysis reveals the Mojave Desert as an optimal location, despite its dwindling size, due to its flatness, availability of sun, and its proximity to major load centers.

Given this location, this study assumes all PV arrays will be facing between southwest and southeast at an elevation of around 30° as this maximizes solar energy production. Shading should also be taken into account, bearing in mind the proximity of local buildings, vegetation and the possible future plans of development or tree growth. Even minor shading can have a significant effect because it is the cell of lowest illumination that determines the current. This is why we set the following characteristics for our model city.

10.2.4 CHARACTERISTICS OF MODEL CITY

Using data from the Census Bureau we estimated that an average American city is composed of 150,000 households. Though more narrowed, city is still a wide term—often composed of mixed residential and commercial space. To further simplify things we decided that our city would be composed solely of residences, much like a suburb close to a metropolitan area. This allowed us to focus our findings on residential consumers, eliminating commercial and industrial electricity use. Furthermore, our model city does not include apartment high rises or town homes.

As for the residences themselves, the average American home is 2349 ft^2 in area [4] and an average Californian residence consumes approximately 6960 kilowatt hours of electricity per year. Following a discussion with Executive Planner, Jim Christensen from Pacificorp we found out that to power a city of this size, 353,350,000ft^2, we would need to generate 120 megawatts of power. In the case of our solar farm, we have to take into account the 7% average loss through the transmission lines. For the sake of conservatism and round numbers, we rounded this 8.4 megawatt loss up to 10 megawatts bringing the total to 130 megawatts.

We will take this model city and utilize it in each of our two case studies. First, we will analyze the requirements of meeting this hypothetical city's needs entirely with residential photovoltaic arrays, with each household equipped with an array of solar panels necessary to meet the household's own electrical needs. For our concentrated dish Stirling engine farm, power will be transmitted from a remote location to the model city. This second case requires the construction of power substation to lower the high voltage being transmitted from the farm into a safer level that can be utilized in homes. The costs of this added piece of capital, along with all the power source-based calculations, will be detailed in the Results

& discussion section under the subtitle, "The case for concentrated Dish Stirling generation".

10.2.5 CONSTRAINING ASSUMPTIONS OF PHOTOVOLTAIC TECHNOLOGY

Given the geographical location of our model city, this study will assume that the array receives five equivalent noontime hours of sun exposure on an average day. This is a slightly conservative estimate; the state's two largest metropolitan areas, Los Angeles and the San Francisco Bay, receive 5.6 and 5.4 noontime hours of sun on the average day, respectively, while other parts of the state receive as much as 7.7 average equivalent noontime hours per day [5]. For simplicity's sake, this analysis also assumes an array generates no electricity outside of noontime hours.

Given that most photovoltaic cells are guaranteed to remain at 80% of starting efficiency after 25 years, as referenced by Black, this analysis will assume that the cells lose generating capacity at a compounded .9% per year. Thus, it will also limit its lifespan to the first 25 years and assume the array possesses no generating capacity afterwards.

Another assumption is the number of times the inverter has to be changed. Over time, the inverter coils wear down and eventually fail. Though there is not yet a consensus over the average life of a photovoltaic array's inverter, estimates range from as little as 4.7 years [6] to longer than the lifespan of the array. For the sake of this analysis, we assume one inverter replacement half way through the lifespan of the array.

Finally this study will only take into account governmental policies that affect the whole state. Particularly, it will consider the 30% Resident Renewable Energy Tax Credit offered by the federal government, and the subsidies offered by the California Solar Initiative. However while several cities and counties offer additional incentives for photovoltaic array installations, these will be ignored for the purposes of this paper [7]. Similarly, this paper will not assume tiered electricity pricing as it is only active in certain parts of California, but it will mention how this may affect our findings.

Research shows that the average efficiency of these cells lies between 13% and 16% [8]. This loss in energy results from thermodynamic efficiency losses (up to 75%), losses in the inverter (10-15%), reflectance losses (~10%), temperature and dust accumulation (10%), and resistive electrical losses (1-3%) [9]. Hence, for the sake of conservatism this study will assume a 13% of cell efficiency. See Additional File 1 for further explanation on how energy is lost and further detail on how this technology works.

10.2.6 CONSTRAINING ASSUMPTIONS OF CONCENTRATED SOLAR PLANT

Given the general location of California this study will set its hypothetical solar plant, for precision's sake, outside the city of Barstow, in the county of San Bernardino. Hence, the plant will be affected by its typical weather of 102°F in the summer, receiving 281 days of sun, and 22 days of precipitation, with annual rainfall of 5 inches.

To determine the value of the land per acre we did the following research. In a ground known as the Mojave Desert Land Trust land prices ranges from $500 [10] to $1,522 [11] per acre depending on the government subsidy. Outside the realm of nature preservation the land prices begin to increase steadily. A survey of available land in Barstow reveals prices of $900 per acre in more rural areas [12] compared to $2,163 [13] and $4,225 [14] per acre closer to the city center of Barstow. Given the requirements of our project we took the average of the three that best meet our land qualities: $500, $900, and $1,522, establishing a cost of $974 per acre.

When it came to defining the lifespan of the plant we found many studies citing a theoretical lifespan ranging from 20 to 30 years. Sean Gallagher, Vice President of Market Strategy & Regulatory Affairs at Tessera Solar, provided a way to think of things more concretely for the sake of our study: the lifetime of a dish Stirling engine is 100,000 hours of run time. Now, given that our dishes will run 12 hours a day we get 100,000/12 = 8,333.33 days of lifetime or 22.83 years. For simplicity's sake and the potential of downtime due to maintenance in the lifetime of the dishes, we set a lifespan of 23 years.

Similarly, over the lifetime of the farm certain routine maintenance would have to be performed. These include a complete washing of the reflective mirrors of each engine eight times a year, as well as engine maintenance once every two years. However, for calculations' sake Sean Gallagher provided another way of determining the costs by calculating maintenance on a kilowatts per hour basis. This is done by defining the amount of grid-ready kilowatt-hours a dish generates in a year and by establishing a cost per kilowatt-hour of electricity generated. This logic shows that the cost of maintenance per kilowatt-hour of electricity generated is less than 2 cents, our case study assumes a cost of 1.8¢ per kilowatt-hour.

Next we define the sale price of energy produced with this technology. Several studies, including Black and Goodward, have quoted a sale price between 6¢ and 8¢ per kilowatt-hour [15], and given that this conservative range is outpaced during peak demand where many areas of California reach 11.33¢ per kilowatt-hour, this study will set the sale price at 8¢.

Finally, we set the initial rate of return (IRR). Given that there are no major doubts related to this technology as it has been tested and proven reliable, but also given that this is quite a large installation and certain speculation remains, as sustained by Leitner, regarding the viability of the project, hence we set an IRR of 20% to help dissuade any doubts of technology risk and help us acquire the necessary level of capital.

10.2.7 SETTING A BENCHMARK AND FORMULATING PRICING ASSUMPTIONS

We will use conventional energy as a benchmark when analyzing both models' benefits and costs. According to the Energy Information Administration (EIA) data from 2009, the average American household consumes 936 kilowatt hours of electricity per month at an average retail price of 10.65¢ per kilowatt hour. This implies that average household consumes $99.70 worth of electricity a month. However, given the specific geographic locality of California, the average household spends $139.56 in electricity a month. Helping to put this into context, the EIA states that the United States produces 4,156,745 (thousand) Megawatt hours (MWh) per year of which 48.5% comes from coal, 21.6% from natural gas, 19.4% from nuclear, 5.8% from hydroelectric sources, 1.6% from oil and 3.1% from others, such as solar and wind energy [16].

Since 1970, as sustained by Black, the retail price of residential electricity in California has risen by an average of 6.7% annually. For our analysis, we assume that this trend will continue for the next 25 years. Under this criterion we expect the price of energy to be 22 cents per kWh by 2015, 42 cents per kWh by 2025 and 80 cents per kWh by 2035. Furthermore, we assume a discount rate of 7%. This rate represents the opportunity cost of investing in a risk free asset plus an extra 2% to accommodate price shocks to electricity. Using these values we estimate the total cost of energy for our model city at a present value of $3,471,909,155. This value represents the aggregate cost of supplying electricity to our 1,044,000,000 kWh town for 23 years. Performing the same calculations for the Sterling Dish farm and taking into consideration the necessary increase in power supplied due to transmission loss, we calculate/find a net present value of $3,763,352,167. These results will be used when comparing the costs of the photovoltaic and Stirling engine models.

10.2.8 SETTING TWO DISCOUNT FACTORS

In order to properly discount for the two technologies we are going to use two separate discount factors. For the home photovoltaic system we will assume the same discount rate we used for discounting energy coming from the national grid, 7%. Here again we assume an initial 5% discount, which measures the opportunity cost of investing in a risk free asset. However, the additional 2% represent the uncertainty in the future price of raw materials such as silicon. For the Stirling engine technology we are going to use a 10% rate. The higher discount rate makes sense in this case due to the higher upfront capital costs and the fact that there is uncertainty due the scale of this endeavor because nothing of this sort has been yet implemented.

10.3 RESULTS AND DISCUSSION

10.3.1 THE CASE FOR DISTRIBUTED PHOTOVOLTAIC GENERATION

Distributed electricity generation is an attractive technology. By reducing or eliminating dependence on the national power grid, the consumer may provide for his or her own electricity demand at essentially zero marginal cost, whilst often recouping the initial capital investment associated with setting up the generation system in future electricity savings and in the value of electricity sold to the power grid.

Photovoltaic solar power is the quintessential distributed generation technology. The power produced by a photovoltaic array scales linearly with the area of the system, so as long as the array produces enough revenue to compensate for the non-generating sunk cost of the system (the inverter, etc.), a photovoltaic array is a sensible economic choice. The only trait required of a location is open, south-facing space for installation when in the northern hemisphere. They have very low maintenance costs, require little attention from their owner, and have a lifespan of 25 years, commensurate with the time horizon of many home-planning decisions—most mortgages are 15 or 30 years.

Unfortunately, commercially available photovoltaic cells remain very expensive for most residential consumers. The key to making photovoltaic arrays a cost-effective alternative to fossil fuels lies in two economic maneuvers on the part of the federal and California state governments.

First, the United States Congress has mandated that a technology and accounting practice called "net metering" be available to all electricity consumers [17]. Under a net metering scheme, any consumer attached to the power grid is given credits for electricity that user produces above his or her own electricity consumption through the use of distributed generation technology. When the consumer is using more electricity than he or she is producing, the electricity is purchased at the normal rate. Then, at the end of the billing period, the credits are subtracted from the bill, and the consumer only owes the utility the difference between the value of the electricity he or she produced and the value of the electricity he or she consumed. Due to net metering, a photovoltaic array allows a consumer to continue to consume electricity, but at a lower price than he or she would purchase that electricity from the local utility company. These savings in future electricity bills add to the value of an installed photovoltaic array.

Second, both federal and state governments provide subsidies for the installation of solar electricity generating systems. The federal government provides a 30% tax credit, for the value of installed residential and commercial photovoltaic systems [18]. This subsidy discounts the taxes of a property owner who installs a photovoltaic system by 30% of the total price of the installed system; for the purposes of our analysis, this is equivalent to the federal government paying 30% of the cost of the photovoltaic array, leaving the remaining 70% to be paid for by state subsidies and the property-owner.

In California, the cost of installing a photovoltaic system is $8.20 per watt of generating capacity, the second lowest in the nation [19]. This cost is increased by the only substantial maintenance cost associated with residential photovoltaic systems: the replacement of the inverter. The price of a solar array inverter is 71.9¢ per watt of generating capacity [20]. Assuming 2% inflation and a 7% discount rate per annum, the present value of this replacement is 38.6¢ per watt. As the price of inverters has been dropping over time, this allowance for inverter replacement will also allow for some routine inverter maintenance in addition to the inverter replacement midway through the 25-year span of this analysis. This increases the cost of each installed watt by 39¢ per watt, bringing the total cost per installed watt of photovoltaic generating capacity to $8.69.

This cost is very high when compared to the cost of grid electricity to residential consumers, at 14.9¢ per kilowatt hour [21]. Given our predetermined assumptions that the array receives five equivalent noontime hours of sun exposure on an average day and has a 25-year lifespan, the lifetime productivity of one watt of photovoltaic generating capability is 45.6 kilowatt hours. Following our other assumptions of a 7% annual discount rate and a 6.7% increase in the cost of electricity, the present value of those generated watts is $5.92; this is only 68.9% of the initial capital investment required to acquire that one watt of generating capacity. However, after the 30% federal tax credit, the array has paid for itself, leaving a 21¢ cost to the consumer per installed watt of generating capacity. This means

the lump-sum rebate given by the state of California through its California Solar Initiative is almost entirely profit for the consumer, leaving the
present value of an installed watt of photovoltaic generation capacity as
substantially positive.

This analysis is complicated by the way California has structured
its rebate. The level of the California Solar Initiative incentive drops as
more solar arrays are installed in the state, and these drops are not applied
uniformly across the state. The current rebate for residential consumers
ranges from $1.10 to $1.90 per watt of installed generating capacity,
depending on the consumer's utility company [22]. This level of subsidy
leads to a profit for the consumer of $0.90 to $1.70 per watt of installed
solar generating capacity; a 10.4% to 19.7% return on investment. In the
future, this rebate is scheduled to drop as low as 20¢ per watt, but even
in this case the present value of each installed watt is almost exactly
zero. However, by the time the California Solar Initiatives have reached
this low level of subsidy, the technology's efficiency and cost will likely
have improved enough for the photovoltaic array to remain a profitable
investment. See Additional Files 2 and 3 for capacity and present-value
calculations.

The meaning of these numbers is more readily grasped by considering
the case of a typical home. The average Californian residence consumes
580 kilowatt hours of electricity per month, or just under two-thirds the
national average. By way of comparison, the average American residence
consumes 936 kWh of electricity monthly [23]. At 14.9¢ per kilowatt hour,
the annual electricity bill of the average Californian residence is $1037.04.
In order to meet fully the annual electricity needs of such a home, it would
need a photovoltaic array capable of capturing an average of 3.81 kilowatt
during the approximately 5 daily noontime hours available to all Califor-
nians During these hours, each square meter of California receives at least
5 kilowatts of power from the sun. Since our constraining assumptions establish that our solar array is 13% efficient and captures no energy outside
noontime sun, a photovoltaic system of 29.3m^2 (302 ft^2) would power the
needs of the average Californian residence.

By comparison, according to ABC New's report, the average American house is 2349 ft^2 in area. Assuming the average house has two sto-
ries of equal size, an array covering only slightly more than one-quarter

of the house's roof will meet the needs of the average American home in California.

At an initial capzital cost of $8.20 per watt, a 3.811 kilowatt system will have a total cost of $31,251. Deducting the 30% federal tax credit reduces the capital cost to $21,875. This cost is further reduced by the California Solar Initiative rebate, which reduces the cost to between $14,635 and $17,683 for the consumer. However, since an array of this size will fully meet the annual needs of the consumer (after annual net metering), the present value of 25 years of electricity bills must be considered. Given our constraining assumptions of a 6.7% annual increase in the price of electricity, a 7% discount rate, and a loss to generating capability of .9% per year, the present value of future electricity savings is $22,581. As these future savings are greater than the out-of-pocket costs to the consumer, installing such an array is a revenue-positive action on the part of the homeowner, earning him or her $4,897 to $7,946. After a single inverter replacement halfway through the 25-year lifetime of the array, this present value is reduced to $3,411 to $6,475. However, this consumer surplus came at a loss to federal and state governments of $13,567 to $16,616. This means each grid-neutral home creates a dead weight loss of $10,157. Of course, this money does not evaporate, it goes to another agent, the photovoltaic array-producing firm. However, it is a loss to the system between consumers and the government.

Even in situations where the present value of future savings on electricity is less than zero, additional incentives remain for homeowners to purchase photovoltaic arrays. The most substantial of these is the boon to home resale value. While estimates vary on the precise level of increase in property value due to the installation of an array, the most common estimate is that decreases in annual operating cost increase home value by a ratio of 20:1. That is to say, an array that made a home grid-neutral would decrease the average California residence's annual electricity bill by $1,037, leading to a $20,741 increase in the property's resale value. The logic underlying this figure is that the annual savings allow the potential homeowner to take a larger mortgage to purchase the home, and the roughly $1,000 saved each year may be put into debt service on a 5% mortgage. A more theoretical analysis would conclude that the maximum increase in property value should equal the present value of remaining future electricity bills at the time of the transfer of ownership of the house. In either case, installing a photovoltaic array is revenue-positive decision for the current owner of the house even if the home is sold the day after the array is installed.

It is important to note that these estimates are somewhat conservative given our constraining assumptions that the array has no value after its 25 year lifespan, that it generates no electricity outside of noontime hours, that the array is in the parts of California that receive the least intense sunlight, and that this study does not take into account tiered electricity pricing since it is only active in some parts of California.

In most cases, tiered pricing on retail electricity will make solar technology more attractive rather than less for most residential settings; in variable cost schemes, the price of electricity tends to be highest during the heat of the day, especially in the summer. At these times, photovoltaic arrays are at their most productive, and are likely to be producing more power than the attached home is consuming. As a result, the array will be pushing electricity onto the grid, generating net-metering credit when electricity is at its highest price. After sunset, when the photovoltaic array is not generating electricity, the residence will be drawing electricity from the grid when the price level is lower.

Of course, the most compelling reason for the widespread adoption of solar electricity generation technology is the reduction of the negative externalities of other sources of electrical power. In particular, the carbon dioxide released by the burning of fossil fuels is understood to be the driving force behind global warming, and is thus a matter of prime concern. For instance, one kilowatt hour of power generation in California correlates to 0.30 kilograms (0.66 pounds) of CO_2 emissions, meaning a grid-neutral photovoltaic array attached to the average California residence initially reduces carbon emissions by 2.1 metric tons per year. Over the 25-year lifespan of the array, accounting for decay in the quality of the land, total CO_2 emissions are reduced by 45.6 metric tons. This equates to 12.2 kilograms of lifetime CO_2 emissions reduced per watt of installed generation capacity. The initial capital cost of these CO_2 emission reductions is 67¢ per kilogram over the lifetime of the array; the federal tax credit is 20¢, the California Solar Initiative rebate is 9¢ to 16¢, and the present value of consumer net revenue per kilogram of reduced CO_2 emissions is 7¢ to 14¢, depending on the level of state subsidy. The economy-wide cost of these reduced emissions is thus 22¢ per kilogram.

This analysis reveals that heavy subsidies from federal and state governments have made photovoltaic arrays a sensible investment for the average residential consumer. If the consumer possesses the available roof space facing in an appropriate direction, a photovoltaic array is a profitable investment yielding 10-20% returns over the lifespan of the array, even after a 7% discount rate, and conservative estimates for the output of the array. Even as subsidies decrease, the increase to a home's property value provide a strong incentive for homeowners to augment their homes with grid-tied photovoltaic arrays. These returns compare particularly favorably to other investments, as they are not subject to taxation; federal law mandates that photovoltaic arrays do not increase property taxes, and the present value of future electricity savings are already post-tax earnings.

10.3.2 THE CASE FOR CONCENTRATED DISH STIRLING GENERATION

The size of our solar farm is determined by the number of Stirling engines needed to power our model city and the manner in which these will be arranged. Each dish Stirling engine produces 25kilowatts on its own [23] given that our model city requires 130 megawatts we would require 5,200 dish Stirling engines. Note that the construction of a solar farm is systematic and allows for each completely installed dish to begin generating electricity prior to the full completion of the farm (see Additional File 4). In this case we have established that the dishes will be installed in sets of 60, each one ramping to productive capacity when installed. Hence, 86 2/3 60-dish installations are required, which we will round up to 87 to cover for extra energy spikes, other engines lost due to maintenance, etc.

Taking conventional estimates into consideration we determine that the plant would required between 780 and 910 acres to accommodate the number of dishes necessary to power our farm sustainably. Note that the traditional means of calculating the dimensions required for a plant, as explained by Gallagher, is to assume 6 to 7 acres per 1MW. To add precision for the sake of later calculations, we will choose 6.5 acres as the requirement per megawatt. Given this we calculate a land requirement of 845 acres. Since our solar farm has been set just outside the city of Barstow in the San Bernardino County, where the expected cost of land is of $974 per acre (See Constraining Assumptions of Dish Stirling System for details), we estimate a cost of $823,030 in order to fully house the required equipment.

According to Sean Gallagher, a 130 megawatt plant size would roughly necessitate 150 construction workers. Due to the nature of the construction we fortunately would not need a specialty construction company or a wealth of engineers. Another bonus of this well-defined, modular construction process is that it allows for 24-hour construction as the optical alignment can take place during the night. See Additional File 4 for details on the construction process.

The construction progresses at a typical speed of one megawatt of generating capacity completely installed and completed per day. Given that each dish represents 25 kilowatts (or 0.025 MW) we get a number of 40 dishes installed per day. This allows for four arrays of 60 to go active every week. Now, assuming completion of 40 dishes a day, and given 5,200 dishes required, the construction process would stretch over 130 days.

There is some difficulty in cost speculation regarding construction as well as parts production related to dish Stirling. This uncertainty stems mostly from the lack of any large-scale plants having been put into commission. Even so we have analyzed the costs associated with similar large-scale construction projects and have come up with the following information.

Port, in a 2005 BusinessWeek article stated that the handcrafted dish itself is a costly monster at $250,000 per rig. Bulk orders, opposed to the one-off tailor made orders, can help lower the costs by roughly $100,000 apiece. Large economies of scale in production promise to lower the cost even further in theory, reaching a sticker price of roughly $80,000 or even $50,000. Further research has shown that the new expectation for "mature price approximation" for the strict production of dish Stirling engines is $1,000 per kilowatt [24], given larger scale production. This number fits well with the cost adjustments achieved with larger installations. Sean Gallagher cited the notion that a 25kilowatt dish Stirling engine costs $75,000 per dish installed—including both the fabrication and installation costs.

This gives a price of $3,000 per kilowatt. This discrepancy of $2,000 can be accounted for by different production cost approximation and the cost of installation. Therefore, given the situation today we estimate a cost of $75,000 for each engine in an ideal production cycle. This implies a cost of $75,000*5,200, equaling $390,000,000 for both dish production and installation.

However, any substantial exploitation of the renewable source will depend on being able to transmit the energy from its source to its final point of usage, in this case, an urban center [25]. Hence, a substation needs to be constructed in order to lower the voltage transmitted by the solar farm. Placing the solar farm roughly one hundred miles from our city means that we need a minimum transmission voltage of 138,000 volts. For the initial calculation we are using a base unit for a 40-megawatt plant and given that these costs are linear we can then adjust for our 130-megawatt solar farm. Assuming high side protection, a circuit breaker will need to be installed which will cost $75,000. Then at the heart of the substation we have the transformer. A 138Kv to 12.5Kv 40 MVA transformer is going to cost $750,000. In addition there is a low side breaker, which recent estimates put at $20,000. Now that we have the large pieces of capital accounted for there is the engineering and parts and pieces need to connect it all together and make it work. A conservative estimate was given of $155,000, which brings our grand total to $1,000,000 for our 40 megawatt substation. Adjusting for our 130 megawatt farm leaves us with a fixed cost of $3,250,000.

As to maintenance costs, these will be calculated on a kilowatts per hour basis, which requires an estimate of the kilowatts per hour received per day. Barstow in San Bernardino County, CA enjoys an average number of 7.587 kWh/m²/day [26]. Knowing that each dish Stirling engine is 38 foot high by 40-foot wide solar concentrator in a dish structure [27], we calculate a surface area of about 111m². Given that this system has an efficiency rating of 31.25% for converting solar thermal heat into grid quality electricity [28], we calculate that out of a total of 7.587 kWh/m²/day hitting Barstow only 2.37 kWh/m²/day will be converted into grid ready electricity. Hence, 96,058.5 kilowatt hours per year can be generated per dish.

Given our established maintenance cost of 1.8¢ per kilowatt hour, we get a maintenance cost of $1,729.1 per dish per year and a total cost of

$8,991,078.67 per year for the 5,200 dishes in the plant. Another way of viewing this, which this study will later use to compare it with photovoltaic cells, is $.069 per watt per year.

Therefore the present value of the maintenance cost over the predetermined lifespan of 23 years, assuming an inflation rate of 2% and a discount rate of 10%, would be 78¢ per watt or $101,855,915 for the whole 130-megawatt plant. See Additional Files 5 and 6 for present value calculations.

From an energy standpoint it appears that the solar farm is primed for commercial success—at least as far as demand is concerned. The solar source delivers very reliable peak power when the sun is shining. This time is ideal for delivery of sunlight, as daytime is the end of the user's peak demand: therefore, peak load equals peak power.

In order to calculate the lifetime profitability of the plant we must take into account the construction costs as well as the fixed costs and upfront capital required for the initial construction. Given the quick nature of the construction process we would need the construction cost, the substation cost, and the cost of the land upfront. In order to acquire this level of capital from investors we must appeal to them with an attractive internal rate of return based on the perception of risk associated with the technology. As stated before, this study has assumed that an IRR of 20% would help dissuade any doubts of technology risk and allow for us to acquire the necessary level of capital.

In order to derive the revenues generated by the Stirling engines technology we used the total energy needed per year for our city: 1,044,000,000 kilowatt hours. Following our constraining assumptions we used a high side estimate of 8¢ per kilowatt-hour, 6.7% increase in electricity per year and 10% discount rate, arriving at a revenue of $1,402,282,942.

Using the above calculations for capital, land and the substation, we arrived to a total fixed cost of $394,073,030. Given that all this money is borrowed upfront we are giving our investors an internal rate of return of 20%. Total interest payment to investors is $78,814,606. Finally, we must account for maintenance cost, which has a present value of $101,855,915. Adding these three numbers together we arrive at a complete lifetime cost of $574,743,551. Given that profits equal revenue minus cost, that is \$1,402,282,942 less \$574,743,551, we arrive at total profits of \$827,539,391.

10.3.3 A DISCUSSION ON POTENTIAL PROBLEMS OF SOLAR TECHNOLOGIES

From the above analysis it is clear that both investments are a revenue positive action. However, some concerns may remain as to its pragmatism given that some potential problems of relying on the sun as a source of energy include seasonality, cloud cover and unpredictability, as well as nightly outages.

Though it may seem obvious there is a lack of sunlight during the evening, a problem that represents an important factor when considering solar energy, the alternative trough and solar tower CSP systems can utilize a hybridization system to combat their nighttime losses. Though less efficient, they utilize natural gas to keep their turbines moving. This is not too large a concern as it utilizes equipment that would otherwise be idle. There have been proposals for the incorporation of a hybrid fossil fuel system into the Dish Stirling system, but it would suffer from lower efficiencies and lose some of its zero emissions appeal. The notion of a mixed fuel system is a disadvantage for the Stirling, as it would need to be an integral part of its design. Regarding photovoltaic cells, although during the nighttime energy would not be produced, during the day the cells should overproduce. The net metering enables the photovoltaic cell to take advantage of electricity from the national grid during times of shortages, but due to its overproduction, stay grid neutral.

Additionally, both photovoltaic and dish Stirling technologies can fall victim to the unpredictability of cloud cover and weather. However, Dish Stirling units have the unique ability to ramp up to full output within seconds. This coupled with their bigger size and ability to track the sun, as explained by Leitner, allows for average output that tracks average radiation levels very well. Still, they suffer similar disadvantages to PV given cloud cover, but they are even worse off given their inability to utilize scattered light.

Solar plant located in Mojave Desert serving Southern California Edison

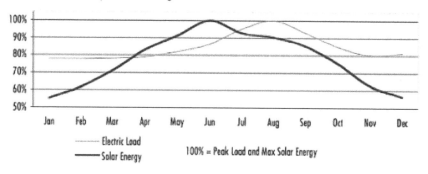

FIGURE 1: Solar resource and electric load in the Mojave Desert.

Lastly, Leitner also explains that in the case of seasonality, clouds and haze reduce output by 20% in December and January. Likewise, shorter days and less direct exposure to sunlight are instrumental in the total output of the Stirling engines. The summer remains the strongest time period for sun collection. However, despite these short falls, solar energy closely matches the electricity consumption cycle of consumers. The energy production is closely correlated with load, increasing in summer when it is most required—air conditioning being a huge factor in this region. The result is almost simply a downward parabola, centered at June (See Figure 1).

10.4 CONCLUSIONS

Given our analysis in the previous sections, we conclude that the dish Stirling system is a superior option. We found that the dish Stirling consumer receives 6.37 dollars per watt while the home photovoltaic system consumer receives between 0.9 and 1.70 dollars per watt. Given these findings, we see that consumers are better off investing in a dish Stirling system. We see a significantly greater return on this technology compared to photovoltaic cells. This, at first, seems odd given that the expenses for

Stirling engines are much greater than that of photovoltaic cells. However, given that the power is ultimately sold back to consumers we were eventually able to realize a profit. Furthermore, once put in the scope of the real world the dish Stirling engine appears to gain more positive moment. For example the feasibility of a solar farm, given its size, can often be brought into question. However, gaining a set of several strong investors seems much more feasible than getting a town of 150,000 households to put photovoltaic cells on their roofs. It is far easier to do the former, which intuitively makes sense. Then there is the issue of efficiencies. We said earlier that the efficiency of photovoltaic cells is between 13-16% while that of the Stirling engine is 31.25%. Based on the higher efficiency of the Stirling engine, it is not difficult to believe that this technology will outperform its rival. However, one thing we did not take into consideration was potential subsidies or grants given for the construction of the farm. These have the potential to drive the costs down even further, increasing the watts per dollar generated, thus further widening the gap between Stirling engines and photovoltaic systems.

If our goal is a reduction of CO_2 emissions, then clearly both methods of electric productions eliminate most CO_2 emissions via reduction of fossil fuel-based production processes. Though there may be some CO_2 emissions during the manufacturing processes these emissions are incredibly small in comparison to the reduction in fossil fuels used.

As to policy implications, given current levels of subsidies and tax credits we found that the home photovoltaic system actually returns a profit to the homeowner. This indicates that these subsidies are too high and the policy is lagging behind new advances in technology. This misallocation could instead be used in the subsidy of dish Stirling farms where it would receive a much higher return.

Stepping away from subsidy policy we must now also consider the environmental impact concerns of dish Stirling construction. The clearing of vast acreages of land poses serious concerns for wildlife habitats as well as water usage issues. One must remember that these farms are located in the Mojave Desert where water is scarce. The Mojave Desert Land Trust was set up to combat the development of these precious ecosystems of the west. This group has taken the initiative to purchase land an incorporate it into preserves, saving animals from possible extinction.

Ultimately, the positive aspects seem to outweigh any minor concerns or potential externalities. The solar farm, and even the less practical decentralized photovoltaic deployments, help alleviate CO_2 emissions as well as maturing renewable energy technology. The major goal is to one day achieve fully sustainable systems, run completely on renewable energy, giving a cheap source of electricity and an all-important source of energy independence.

REFERENCES

1. Port O: Solar Power's New Hot Spot. [http:/ / www.businessweek.com/ technology/ content/ aug2005/ tc20050819_0041_tc024.htm]
2. Leitner A: Fuel from the Sky: Solar Power's Potential for Western Energy Supply. National Renewable Energy Laboratory; 2002.
3. Marathon Capital LLC: Case Study on Solar Power Financing. 2008.
4. ABCNews: America's Homes Get Bigger and Better. [http://abcnews.go.com/GMA/ Moms/story?id=1445039] 2005. Accessed October 27, 2009
5. Black A: Economics of Solar Electric Systems for Consumers, Payback and other Financial Tests. [http://www.ongrid.net/papers/PaybackOnSolarSERG.pdf] San Jose; 2009.
6. Bonn RH: Developing a "Next Generation" PV Inverter. In Proceedings of the Photovoltaic Specialists Conference: 19-24 May 2002. Albuquerque;
7. North Carolina Solar Center, DSIRE: Database of State Incentives for Renewables & Efficiency. [http://dsireusa.org/] 2009.
8. Chartered Institution of Building Services Engineers Staff: Understanding Building Integrated Photovoltaics. Chartered Institution of Building Services Engineers; 2000.
9. Harwell: A Study of the feasibility of photovoltaic modules as a commercial building cladding component. In Report no. S/P2/00131/REP. Energy Technology Support Unit; 1993.
10. Karl N: The Mojave Desert Land Trust Reaches 10,000 Acre Milestone. [http:// www.mojavedesertlandtrust.org/pdf/Pr.Rel.10KAcres.Jan09.pdf] Mojave Desert Land Trust; 2009.
11. Sall C: The Mojave Desert Land Trust: Working to keep the Mojave...Desert. [http:// www.mojavedesertlandtrust.org/newsletters/MDLT-Newsletter12-06.pdf] The Mojave Desert Land Trust; 2006.
12. Weichert Realtors: 160 Acres, Barstow CA 92347: Lots of land in Barstow, CA. [http://www.weichert.com/26385121/?cityid=2965&mls=146&ptypeid=3%2c30]
13. Weichert Realtors: 45.74 Acres, Barstow CA 92365: Lots of land in Barstow, CA. Accessed on http://www.weichert.com/26123353/?cityid=2965&mls=146&ptypeid =3%2c30

14. Weichert Realtors: 20 Acres, Barstow CA 92311: Lots of land in Barstow, CA. [http://www.weichert.com/21560578/?cityid=2965&mls=146&ptypeid=3%2c30]
15. Goodward J, MacBride A, Rigdon C, Staley BC: Juice by Concentrate: Reducing Emissions with Concentrating Solar Thermal Power. World Resources Institute; 2009.
16. U.S Energy Information Administration: Form EIA-923. [http://www.eia.gov/cneaf/electricity/page/eia423.html] 2009.
17. The Energy Policy Act of 2005 §1251 [http:/ / www.eia.gov/ oiaf/ aeo/ otheranalysis/ aeo_2006analysispapers/ epa2005_summary.html]
18. North Carolina Solar Center, DSIRE: Database of State Incentives for Renewables & Efficiency. [http://dsireusa.org/] 2009.
19. Wiser R, Barbose G, Peterman C, Dargouth N: Tracking the Sun: The Installed Cost of Photovoltaics in the U.S. from 1998-2008. Lawrence Berkeley National Laboratory; Environmental Energy Technologies Division; Berkeley; 2009.
20. Solar Buzz, LLC: Inverter Price Environment. Retrieved 11 November; 2009. http://www.solarbuzz.com/inverterprices.htm
21. U.S. Department of Energy: U.S. Average Monthly Bill by Sector, Census Division, and State. Retrieved 28 November 2009 from Energy Information Administration http://www.eia.doe.gov/cneaf/electricity/esr/table5.xls
22. California Solar Initiative - Statewide Trigger Point Tracker [http://www.csi-trigger.com/] 2009.
23. Schlaich Bergermannund and Partners Structural Consulting Engineers: EuroDish – Stirling System Description. 2001.
24. ACEEE, American Council for an Energy-Efficient Economy: Emerging Technologies & Practices. [http://www.aceee.org/topics/emerging-technologies-and-practices] 2004.
25. Whittington HW: Electricity generation: options for reduction in carbon emissions. Philos Transact A Math Phys Eng Sci 2002, 360(1797):1653-1668. PubMed Abstract | Publisher Full Text
26. National Renewable Energy Laboratory's Solar Power Prospector [http://mercator.nrel.gov/csp/] 2009.
27. California Solar Initiative - Statewide Trigger Point Tracker [http://www.csi-trigger.com/] 2009.
28. Tessera Solar [http://tesserasolar.com/international/award-winner.htm] 2009.

There are several supplemental files that are not available in this version of the article. To view this additional information, please use the citation on the first page of this chapter.

PART V

PUBLIC PERCEPTIONS OF SOLAR ENERGY

CHAPTER 11

EXCERPT FROM: COMMUNITY RESPONSE TO CONCENTRATING SOLAR POWER IN THE SAN LUIS VALLEY

B.C. FARHAR, L.M. HUNTER, T.M. KIRKLAND, AND K.J. TIERNEY

11.1 PERCEIVED ADVANTAGES

11.1.1 INTRODUCTION

Stakeholder informants were asked about their organizations' view on the advantages and disadvantages of renewable energy generally and of CSP in the San Luis Valley. The 396 pages of text were content-analyzed to discern discrete mentions of perceived advantages of renewable energy generally in the context of open-ended questioning; these are considered "responses." Five broad categories of advantages were developed from the comments themselves: (1) economic, (2) environmental, (3) technological, (4) social-psychological, and (5) policy and regulatory. Under each

This chapter was originally published by the U.S. Department of Energy. U.S. Department of Energy, National Renewable Energy Laboratory, Community Response to Concentrating Solar Power in the San Luis Valley, by B.C. Farhar, L.M. Hunter, T.M. Kirkland, and K.J. Tierney, NREL/SR-550-48041 (June 2010). http://www.nrel.gov/csp/pdfs/48041.pdf (accessed 30 June 2014). The selection excerpted here is from pp 14–56 of the original pdf.

of these broad categories, subcategories were defined. Each response was coded by category and subcategory.

Table 1 shows that 231 discrete responses mentioned advantages of renewable energy generally and CSP specifically were identified. The table is organized in terms of decreasing frequency of mention (or responses) of the different types of advantages, with economic advantages receiving the most emphasis in the data set overall (44%), followed by environmental advantages (26%), technological advantages (19%), social-psychological advantages (9%), and policy and regulatory advantages (2%).

Overall economic advantages are mentioned far more frequently than are environmental or other types of advantages by both types of stakeholders. Stakeholder interviewees inside the SLV mentioned economic advantages of CSP specifically in the Valley almost twice as often as they mentioned economic advantages of renewable energy generally, suggesting that local economic advantages of a proposed CSP facility weighs more in locals' positions toward the facility than other considerations.

TABLE 1: Types and Percentage Distribution of Responses about Perceived Advantages from Stakeholders within and outside of the San Luis Valley (n=231 comments)

Types of Advantages	Renewable Energy Generally		CSP Specifically		
	% Responses Within SLV	% Outside SLV	% Responses Within SLV	% Outside SLV	% Totals
Economic	27	38	53	52	44
Environmental	40	34	20	15	26
Technological	25	15	11	26	19
Social-psychological	4	6	16	7	9
Policy and regulatory	4	7	--	--	2
	100% (n=48)	100% (n=53)	100% (n=64)	100% (n=66)	100% (n=231)

Stakeholders outside the Valley more often mentioned "technological" advantages than did those in the Valley, suggesting their higher level of technical sophistication about CSP. Third, environmental advantages of both renewable energy generally and of CSP specifically were mentioned somewhat more frequently by Valley stakeholders than by outsiders. Fourth, Valley stakeholders mentioned "social-psychological" advantages more frequently than did those outside the Valley. Each of these the categories of advantages is discussed and exemplified in more detail below.

11.1.2 ECONOMIC ADVANTAGES

Economic advantages identified tended to focus on the potential economic impacts of CSP within the Valley, rather than on effects of renewable energy generally. The 44% of discrete responses categorized as economic (n=67) fell into six somewhat overlapping categories of advantages, discussed below. Most frequently mentioned were potential improvements to the local SLV economy from the siting of a CSP facility there. Two quotes from Valley stakeholders:

> ...it'll happen. For everyone [but farmers], this is a no-brainer. If we just had the wherewithal as a little rural community to make this happen and to be sure that the SLV reaps the rewards...how can we make sure that the SLV reaps the rewards? Because someone could build a solar plant and all the money leaves the local economy...we'll, we've done good for our planet, but how can we make this benefit our own economic development?

> ...our intent to have this valley 100% renewable by the year 2010. And that could happen in an hour if Tri-State, Xcel, and Rural Electric all agreed that they could get mutual benefit because we could actually purchase enough [renewable energy credits] to have the Valley 100% renewable based on existing energy...

Proportionally more stakeholders outside the Valley than within it referred to economic benefits for the people living in the SLV; it was one of several benefits of CSP that they cited to describe the benefits of siting a CSP facility there.

...the new energy economy has been a huge boom to our state with respect to job creation and economic development at a time when we really need it the most.

They use a local resource so they create economic activity in the Valley . . . It adds another engine to the Valley's economy. The Valley is pretty much agricultural as I understand it, so you bring in a very clean other business.

....having the ability to create a new sustainable energy industry, also looking at the opportunity that we could generate as much as $2 billion in new investment in rural communities which are so desperately in need of....

On the other hand, Valley stakeholders stressed the idea that renewable energy generally creates jobs and that a local CSP facility could lead to the location of a manufacturing facility in the Valley. This notion was circulating fairly widely within the Valley at the time of the interviews, and appeared to be based on statements made by the CSP industry.

And it would create a large number of construction jobs—about 1,000-2,000 over a two-year period) and a smaller number of maintenance jobs, say 60 to 70 full-time jobs—permanent, good-paying jobs for the 30- to 50-year life of the plant....
A local informant said:

It's clean, we have a lot of sun, there are no negative effects and it makes some jobs—and there may be manufacturing jobs—if this were a big site for solar, it could lead to the manufacturing of solar components. That would improve the economy. It would be good for some businesses—contracting with them to place on rooftops, power their business and sell excess to the SLV REC [Rural Energy Cooperative] or Xcel. It's an advantage for business.

Another economic advantage of renewable energy generally seen as crucial (but not specifically mentioned as an advantage of CSP) is that it could stabilize energy prices at a time of rapidly increasing prices of coal and natural gas.

> *Fuel price stability would probably be the single most important facet with respect to CSP...the Governor's Climate Action Plan... calls for a 20% reduction in statewide greenhouse gas emissions by 2020 and an 80% reduction by 2050....it is important to note [that] the current Climate Action Plan models a 200-MW concentrated solar power facility somewhere in Colorado.*

Two other economic advantages were mentioned: (1) providing tax benefits to local counties (which was mentioned more frequently by stakeholders outside the Valley [six mentions] than by ones within the Valley [one mention]) and (2) trapping the dollars in the state's economy that are now sent out of Colorado. In fact, one Valley interviewee commented that there should be a severance tax for exporting solar electricity from the Valley.

> *Do you recall the Federal Mineral Lease, or the Severance Tax, things that are going on with all of that? Did you hear the Western slope and Grand Junction and that area talk about their severing the minerals from our...we should get some of the benefit of that here, you've heard them say that?*

Interviewer: *Yes.*

Stakeholder*: While solar is not a severable mineral, locating your things on our vistas is use of our resource. Should there be a severance tax and could that tax benefit these communities?*

Interviewer: *For economic development?*

Stakeholder: *Yeah, we've got the poorest school districts in the state, I mean on and on. Could the social gain be...we have this resource here, and people from the outside want it, they want our sunshine that's produced in our valley, could we tax for their use of our sunshine to benefit communities located here?"*

Although the supporters of CSP development would probably not agree, the prospect of extra local tax benefits tailored to the solar resource combined with the clear air and flat land that would help mitigate the poverty in the San Luis Valley would very likely be welcome to the Valley community.

11.1.3 ENVIRONMENTAL ADVANTAGES

More than a quarter of the responses about the advantages (26%) addressed environmental advantages (n=60), mostly of renewable energy generally. Most frequently mentioned were that renewable energy, and CSP specifically, puts excellent San Luis Valley solar and natural resources, including lands, to best use (25 responses). Following are quotes from SLV interviewees.

> We have 8,000 square miles, a lot of which is chico brush and land that is not being utilized, so there is plenty of room here.

> I think the land owners who have owned....it's just not good farmland. It's chico brush and either maybe they don't own the water or the soil is not right because there is a lot of alkaline in the valley....so, like where they put that SunEdison...maybe that is the highest best use for that. So the people who sell that land, that may be their only opportunity to sell that land.

> If you do some due diligence you can pick areas of the Valley that are fairly rocky, areas were farming is not done because soil isn't good for farming potatoes. Other areas aren't producing much of anything—those areas would be the priority areas for solar panels. [The interviewee was talking about the 8.2 MW Sun Edison concentrating PV plant—this confusion between CSP and the PV plant existed among several SLV stakeholders.]

Another response was:

If you've got property that received the same amount of sun as the irrigated acreage that provides potatoes for food, you wouldn't take one in favor of losing the other, but you would select the best one for the use. . . I suppose that says this makes more sense to sort of take the crops out of projection if you choose fallow land. It's the highest cash use.

A second type of environmental advantage cited (15 mentions) is that renewable energy reduces greenhouse gas emissions, addresses climate change, and helps to "save the planet." It can probably be assumed that this type of advantage is thought to apply to CSP as well, although no responses cited this advantage relative to CSP in the Valley specifically. Quotes from within the SLV exemplifying this type of comment are as follows:

Whether you believe in climate change or not, the climate's differ-ent than it used to be for whatever reason. The fact is that we need to address that. I think renewable energy is important [in address-ing climate change].

The biggest advantage is that it is renewable, not a finite source of energy. Reduction of greenhouse gas emissions is a benefit to renewable energy development—carbon reduction.

Quotes from outside the Valley on this point include:

I think in order of importance, probably climate change and greenhouse gas emission reduction targets, secondarily the eco-nomic development piece, and perhaps third the fuel price stabil-ity and long-term cost reduction.

Quantified—1 billion tons of carbon reductions.

The third type of environmental advantage is the reference to renew-able energy as "clean energy" and "good for the environment." Thirteen discrete responses, mostly from SLV interviewees, referred to clean energy, but only one relative to CSP. Example quotes are as follows:

> *Renewable energy is good simply because it is renewable...it is popular to develop renewable energy because of environmental reasons, like clean air and water.*

Other environmental advantages identified were that renewable energy uses less water than other forms of energy (two responses—one inside and the other outside the Valley), that it helps wildlife (two SLV responses), and that it won't disrupt the vistas (two outside responses).

11.1.4 TECHNOLOGICAL ADVANTAGES

Emerging from the content analysis was a set of stated advantages that were classified as "technological" advantages, comprising 19% of the responses. SLV stakeholders mentioned at least 17 of these and stakeholders outside the Valley mentioned 25 such responses.

Twenty responses mentioned that renewable energy improves the security and diversity of power supplies and reduces dependence on imported energy. One Valley stakeholder put it this way:

> *[What's important] is the ability to attach and I think what it is called is...they've got to be able to integrate with our system through a substation to get it out, and right now there is a 230, actually it's a double-circuit, 230 transmission line planned in the San Luis Valley to Walsenburg, and it is a combination of Xcel and Rural Electric and Tri-State that are going to put that transmission line in.*

Interviewer: *Is that for sure going to...?*

Stakeholder: *Oh, yeah. One way or another. And it should be.... the SLV doesn't put any money in, it is just for the Valley system,*

*and Xcel and Tri-State pay for it. It is about an $80 million proj-
ect. It should be completed by 2012 or 2013.*

Interviewer: *And why are they doing this?*

Stakeholder: *Because we are running out of power in the valley.
The last transmission line built in San Luis Valley was 35 years
ago.*

One stakeholder from outside the Valley described the advantages of
renewable energy in a different way. This interviewee stressed the security
value of distributed generation, as shown in the following quote:

*We talk about what we're doing today as the alternative model and
that what we'd be doing in the future is going to something traditional, I would say the centralized model is the alternative model.
For the life of the planet, we have dealt with distributed generation and energy for all but about 70 years of the life of this planet.*

*Integration of the distributed and larger grid can complement the
central system with energy efficiency and renewable energy, instead of having another 25 or 30 coal plants in the queue in this
country; it should be classified as preposterous, when we know
that with energy efficiency, conservation, and the distributed model, we can take this centralized system that we've built and is in
play and is working well for us, and complementing it and meeting
our future demands with a much more distributed model.*

*It's the smart grid that is the glue or provides the way to accom-
modate distributed and central station. Once we got the next advances in storage, then we really have a system...*

*Advantages of...It's more resilient. If you look at it from a national
security perspective, and you quantify and weigh national security, economic security, and environmental benefits, this model is
what maximizes those three things.*

A well-informed stakeholder from outside the Valley talked about the prospect of CSP development and the need for a transmission line to enhance the security of the Valley, which has been served by only one transmission line coming in from the north.

> *It was a cooperative transmission line across La Veta Pass into the Valley. It goes across some private land, some state land, and then it also might go partially along a highway corridor, so you're bringing it across the pass. What it is is it is a double, I think it is a 230-KV line that will be a shared project from Tri-State and Xcel. One, they have to have a loop system right now into this valley. Their energy into the valley is very insecure because if there were a fire, these guys are toast. There wouldn't be any electricity for quite a while because they are very vulnerable the way the whole transmission network is in the San Luis Valley.... it's for the security of the region.*

Another outside stakeholder put it this way:

> *It's flipped. Energy independence probably takes precedence over clean energy right now, and that's the whole Middle East thing. So the priorities kind of swapped. In fact if you listen to Obama's speech he talks about eliminating our dependence on Middle Eastern oil. I mean he says Middle Eastern which is key, by, in ten years, and that's all security.*

Clearly, the energy security implications of renewable energy were described from differing perspectives, but similar basic themes emerged about increasing U.S. self-sufficiency through decreasing foreign imports, and increasing Colorado self-sufficiency by producing Colorado power (not importing coal, for example).

A second type of technological advantage was classified as CSP provides dispatchable energy and is a building block toward baseload electricity. Outside stakeholders (n=13) made more comments about this than did Valley stakeholders (n=5).

CSP has two distinct advantages over PV right now. I think costs are roughly equivalent on an energy basis. However, CSP does not suffer the short-term intermittency that PV does.... with the SunEdison facility, there is more volatility with generation...CSP avoids this because it is a thermal cycle. It is essentially a steam turbine and it does not react to individual clouds the way PV does, and it has the ability to add storage inherent in the system . . . the advantage of thermal storage is that with 100% reliability it meets peak day loads.

11.1.5 SOCIAL-PSYCHOLOGICAL ADVANTAGES

One of the most intriguing perceived advantages identified by Valley stakeholders was the discovery that they see the potential for social-psychological advantages. A verbatim sequence of one of the SLV stakeholder interviews illustrates this point:

The sun is readily abundant in the valley. This is the land of cool sunshine. It's colder than heck in the winter but we do get a lot of sun, so...there are actually some advantages I understand from being at this altitude and this climate for concentrated solar because we don't overheat such as you would say in Death Valley. That's my understanding, I don't know if that is totally accurate, but I see that as a benefit. I think....and this is a real intangible, I don't know if you know this, but the San Luis Valley has five of the poorest counties in the state and this is something to be proud of.

Interviewer: *Something cutting-edge....*

Stakeholder: *Yeah, and, well, it puts a different face on the San Luis Valley as far as people on the Front Range come through here and don't understand the beauty of the place because either you love it here or you hate it, but a lot of people from the metro areas are like, oh my God, how can you live in a place like this, and a lot of people who live here don't appreciate what they have, so I* *think changing our perception of ourselves....I mean this could be....and that's why I have a concern about....if they come and build everything and then take everything out of the valley, if they could just come and build everything and share some of that...it's kind of like Los Alamos, you're familiar with Los Alamos? Well, it has the highest concentration of Ph.D.'s in the country, so if some of that technology comes here and some of the people, some of the subject matter experts, and some of the industry associated with it, for example the manufacturing of the actual...whether it's instrumentation, maybe it's just like a silicon chip, I don't know what it is, if they bring industry with it that is associated with it and turn this into....it can't be the Silicon Valley, but more value-added than putting the stuff here and then taking everything out.*

Interviewer: *...somehow there is a local involvement in that there is some kind of manufacturing capability and interest that stays in the valley. Is that the idea?*

Stakeholder: *Yes. That they're committed...that they become a valued business and they are committed to the sustainability of our valley. I don't think the majority of the citizens in this valley want to change. We don't want to be another metro center. We don't want to be an Aspen or a...we'll never be that, but we want to sustain the quality of life...*

Interviewer: *Why not?*

Stakeholder: *Because the quality of life that is here is that we are not Aspen, we're not Vail, we're not Denver, we're not Boulder—that's the quality that needs to be embraced by a business that comes here and understood by a business that comes here. One of the unique things—and this is part of our mission, to enhance and maintain the unique culture and heritage of the Valley because of the strong Hispanic influence, the Native American influence, the Mormon influence—the history of the valley needs to be under-*

stood by companies that come because the majority of the people who live here like it the way it is.

Another Valley stakeholder put it this way:

If you look at demographics of the Mississippi River Delta that compete with the San Luis Valley for a poor economic situation, the valley itself is a distressed economic area. What would be the impact to the rest of the world, the rest of the United States, the rest of Colorado, if the poor San Luis Valley pulled itself up by the bootstraps to where it becomes the first renewable valley in America? There would be entire vital reach in the ecosystem, the whole dynamic of its own future is suddenly 100% renewable energy. Now that's an opportunity to erase at least one level of myth of what you are as opposed to what you aren't. It is a way for people to socially elevate themselves to be able to walk into any store....

Interviewer: *A source of pride.*

Stakeholder: *It is a source of pride. I mean a little of it is that I don't like the rah-rah, go purple and white, fight-fight story...but there are aspects about being able to tell folks that with existing infrastructure, existing people, existing talent, existing local money, that you became the first 100% renewable bio-region in the nation, that the headwaters of the Rio Grande is 100% renewable.*

11.1.6 POLICY AND REGULATORY ADVANTAGES

The final category of perceived advantages was the recognition that the development of renewable energy generally helps Xcel Energy meet its renewable portfolio standards (RPS) requirements, with two responses from the Valley and four from outside the Valley on this point. An illustrative quote from outside the Valley:

I think Amendment 37 was the primary reason that [Xcel Energy] started acquiring solar resources...Amendment 37 tells each

Qualified Retail Utility in Colorado how much of its retail sales have to come from renewable energies, and there is a carve-out within that for solar technologies subject to a retail rate impact cap. If it was going to cost too much money to acquire those resources, then the utility doesn't have to meet the energy requirements.

Amendment 37 was the first citizen-initiated, statewide RPS in the nation; many SLV citizens support the RPS and want it to be met, but did not mention specific advantages to the Valley from it.

11.2 PERCEIVED DISADVANTAGES

11.2.1 INTRODUCTION

The field notes were also content-analyzed to discern discrete mentions of perceived disadvantages of renewable energy generally and of CSP development specifically. The responses on disadvantages lent themselves to categorization in the same categories as the responses on perceived advantages: (1) economic, (2) environmental, (3) technological, (4) social-psychological and (5) policy and regulatory. Again, under each of these broad categories, subcategories were defined. Each response was coded by category and subcategory.

Table 2 shows that 158 responses mentioned disadvantages of renewable energy generally and CSP specifically. The table is organized in terms of decreasing frequency of mention (or responses) of the different types of disadvantages, with environmental disadvantages receiving the most emphasis in the data set overall (42% of the responses), followed by economic disadvantages (23%), technological disadvantages (19%), social-psychological disadvantages (11%), and policy and regulatory disadvantages (4%). One outside stakeholder said there were no disadvantages of CSP siting at all.

Comments on environmental disadvantages occurred more frequently in the interviews than other types of disadvantages and represented the most important area of concern for both types of interviewees (within Valley n=37; outside Valley n=30). Responses from the Valley cited economic disadvantages for renewable energy generally and CSP development specifically (n=26) more frequently than responses from outside the Valley (n=10).

TABLE 2: Types and Percentage Distribution of Responses about Perceived Disadvantages from Stakeholders within and outside of the San Luis Valley (n=158 comments)

Types of Disadvantages	Renewable Energy Generally		CSP Specifically		
	% Responses Within SLV	% Outside SLV	% Responses Within SLV	% Outside SLV	% Totals
Environmental	24	40	41	48	42
Economic	38	13	27	16	23
Technological	38	20	13	26	19
Social-psychological	--	20	13	8	11
Policy and regulatory	--	7	6	2	4
	100% (n=8)	100% (n=15)	100% (n=85)	100% (n=50)	100% (n=158)

11.2.2 ENVIRONMENTAL DISADVANTAGES

The most frequently mentioned disadvantages of CSP development were environmental issues. The two major perceived environmental disadvantages are (1) water constraints and the (2) land-use intensity of CSP.

11.2.2.1 WATER CONSTRAINTS

The most frequently occurring environmental comments concerned water availability by stakeholders within (n=11) and outside the Valley (n=12). The following two quotes are illustrative.

> *CSP is problematic. It is very problematic in terms of the region's ability to conceptualize what it is going to look like and what it*

*is going to take, that water is paramount. You have a hydrology
unlike any other place in the world and you have citizens in this
region who are so committed to their water and using their water
carefully, using this water...it's a closed basin...aquifer, part of it
is a closed basin and then above it is an open aquifer that is what
provides the water for the irrigation. It's an extraordinary bit of
hydrology here...it's the most contentious issue here. That's the
problem with CSP. It requires water. You have some of the most
unbelievable water [battles over water]...these are people who
are more incented by their environment, their air quality, their
water quality, and their quality of life than they are incentives by
having industry come in...maybe that would change if they got a
taste of it, but right now, anything that threatens their water has
them quite concerned.*

*Developers need to secure access to water—it depends on cool-
ing water, cooling tower—evaporates water to cool it. CSP needs
700 gallons of water per MWh. This could be an issue in the SLV,
but it's not a show-stopper. Fans can be used to air dry the plant,
but they consume electricity; this would be an economic decision
based on the cost of water, the availability of water, and a pro
forma financial model. Technically, at least, it's feasible...every-
body knows the SLV is an ideal location; farming is on the decline
because the water table is lower; people are trying to figure out
how to make money. They want to tie up both land and water...
water is in shorter supply than the land.*

Another outside stakeholder said:

*You shouldn't do CSP without water (just like you shouldn't do
coal without water), because it ruins the economics.*

The integrated analysis of energy and water resources is gaining more
attention in research circles as it is increasingly understood that developing energy resources depends on water, and developing water resources depends on energy. Water has historically been and remains a contentious issue

in the Valley, and knowledgeable interviewees acknowledged this. Indeed, a stakeholder analysis of the correlation of CSP and water stakeholders and issues in the SLV should be the subject of separate analyses, and other analyses are needed on the regulatory implications of CSP and water.

Complicated and difficult-to-understand water issues are involved. According to some Valley residents, the SLV has north and south aquifers. Saguache is a closed basin with a three-layered physical structure. At the top layer is surface water; the layer underneath is hardpan; the aquifer lies below. Local residents said that because the layers are relatively impermeable the surface water does not recharge the aquifer. Natural artesian wells spring up from the aquifer through the hardpan. According to some local residents, water is also pumped out to New Mexico to meet water compact agreements. One of the interviewees said that the water priority system is being circumvented by federal subsidies. The highest priority was said to be surface water rights. With more drilling, the surface water was said to be drying up, so the system now is one in which people have to buy and pay for the ability to drill based upon the amount of surface water they own. Based upon their ability to own or buy surface water rights, some irrigating farmers will stay in irrigation and continue pumping. Others will refuse to pay for surface water rights, in which case federal subsidies compensate them for taking their land out of irrigation.

11.2.2.2 LAND-USE INTENSITY

NREL estimates that CSP technology requires five to ten acres of land per one MW of electric capacity. If this is the case, then 500 acres to 1,000 acres of land would be needed for a 100-MW facility. To be cost-effective, a CSP plant has to be a minimum of 100 MW in size and is probably limited at 250-300 MW per facility. One Valley stakeholder said that one of the CSP developers had said publicly that the company was looking at 5,000 contiguous acres on which to site a facility.

Following are comments from the interviews on the perceived land-use intensity of CSP.

Solar still takes up space in our area.

[Solar energy] is an intensive use of land...

...footprint is a disadvantage in general for renewables...the footprint interacts with habitat loss because it requires such a large footprint, so that it's a significant concern.

There are rumors out there. Large corporations from California, from Australia, from I think I've heard Europe, that are talking to groups of people here about acquiring land for concentrated solar.... It's a small community so you hear all of this stuff going on. You don't know what to believe.

Another point a stakeholder made is that there are sensitive wilderness areas in the San Luis Valley. The example mentioned was Saddleback Mountain, reaching 7,956 feet, located in Conejos County and used for hiking.

Developers could see threat in land speculators who are driving up the cost of land or tying up the land. Developers could see threat in "locking up" land to prevent development of CSP projects, although this information must be closely held in the Valley as interviewees did not mention any specifics.

Stakeholders varied in defining how much land the Valley comprises. One said that 5,000 acres is one-third of the square footage of the Valley. Another said that 95% of the land is federally managed. A third mentioned that each of the six counties in the SLV has a different pattern of land ownership. Another stakeholder said, "...well there are about 5 million acres in the San Luis Valley, and people say, you're going to use 5,000 acres with solar panels, and I say, but you have 5 million acres."

11.2.3 ECONOMIC DISADVANTAGES

The most frequently mentioned economic disadvantage of CSP development was that a boom-bust cycle could result. The cycle would involve the hiring of workers to build the facility with the concomitant economic impacts such as increased demand for housing and increased prices in general, crowding, and demand on public services, followed by a decline in the number of jobs once the facility is built. This boom-bust cycle in energy development has been well-documented in the literature. One comment from an interviewee outside the Valley related to this is as follows:

> *I think the biggest concern for local residents is something that can happen on a national basis anywhere, whether it is a wind development or a coal plant or a Wal-Mart, is that locals want to know that they will be benefitting, that there is a long-term benefit to them, that there are potential employment opportunities, and that essentially a company is not going to come in and develop a project and then abscond once the rewards have been taken, that the equipment will be renewed and so forth and so on.*

Many Valley interviewees used the 8.2-MW SunEdison concentrated PV facility near Mosca as a point of reference in discussing the expected impacts of a potential CSP facility.One said:

> *They had temporary jobs to install the solar facility. There were 35 to 38 people hired. It paid well. There were 300 at the job fair, and there was a lot of interest because it was good pay—$14 to more than $20 per hour. These were laborers, electricians, welders, and equipment operators. [The only job since the plant was built] is the main supervisor. I don't know of any continuous employment from the SunEdison plant.*

This interviewee said that when the oil and gas industry came into Parachute, Colorado, they created a boom of jobs paying good money, so the rent on homes there had been increasing. Then the local people couldn't afford the higher rent, he said, so although it was good for businesses like hotels while the boom was going on, the overall economic situation was not good for the local residents. The boom cycle was not sustainable.

11.2.4 TECHNOLOGICAL DISADVANTAGES

Responses in this category stressed the problem that, for a large CSP facility to be sited in the Valley, transmission is needed because the Valley is remote from electricity load centers. Three illustrative quotes follow.

> *One disadvantage is that the electric transmission infrastructure in the Valley is severely limited so infrastructure would have to be built.*

> *The challenge of connecting to the existing grid...*

> *The direct drawback would be how are you going to export it? It goes back to bringing in new transmission lines.*

Another technological disadvantage was that the efficacy of CSP was still in question. CSP was perceived by some not to be a proven technology, not yet ready to be commercialized, and that it will take time for it to be a reliable technology. Two illustrative quotes follow.

> *With the solar thermal with storage carve-out, it was an attempt to push that technology along, to see how it works in Colorado. Colorado has a good solar resource but we've got extreme winter cold temperatures and we have got concerns about how solar thermal technologies will perform in those extreme winter conditions.*

> *Current solar thermal plants have to be maybe 100 MW in size before they become cost-efficient.*

In addition, some comments from within and beyond the Valley questioned the efficacy of CSP as not yet a proven technology or not yet at commercial status. For example, one Valley stakeholder said:

> *...at this meeting they said concentrated solar hasn't been proven. That's what [one of the Valley experts] thinks, and his opinion*

matters...this technology is pretty much all in an experimental phrase. Yeah, some places have had it for awhile, but they are still doing a lot of experimentation.

The intermittency of solar power was also mentioned several times (n=12 comments). An outside stakeholder said, for example, that a CSP plant would probably have a lower "availability factor" than a coal plant.... "the sun goes down, but the storage will supplement some of that, but if you pushed storage to the point where it was an around-the-clock plant, you'd probably be out of the economics."

11.2.5 SOCIAL-PSYCHOLOGICAL DISADVANTAGES

Fifteen comments, most by Valley stakeholders, fell into the category termed "social-psychological" disadvantages. There was concern that a CSP facility would mean inequitable benefits—"short-term benefit for the very few"; CSP would be difficult to conceptualize and people would not understand the impacts. One Valley stakeholder said: "I think they're favorable to solar, but they don't know what concentrated solar is. I would say the majority of people, even a super-majority, don't know what that is."

Another point mentioned several times was the fear that renewable energy is "too green" and that it could be risky to allow it to come into the Valley. One interviewee put it this way:

...we'll have folks, elected officials, who will say, 'well this concept is too green for me' with the idea that if I have to agree with every environmental condition of every group that comes in the door, so there is almost a piece that if I let down the guard protecting the history of how we've done things in the past and instead pick up the idea that by unlocking the door, I'll get more.... If you decide too fast, you can be judged the fool, but if you judge too slow it is not often that you become the hero . . . and we are trying to answer the question, is it safe? Not the technology side, but that they would be judged as making inappropriate decisions...there's the risk. If you say let's go build it and it's financed...but in the end

it doesn't produce power or some new technology comes out that is better or if someone changes the laws...well, they'd say those guys didn't think that through very well.

There is also the sense that it is difficult for the Valley people to deal with Xcel Energy.

There are collaborations [between the rural electric cooperative and Xcel Energy] as far as the transmission lines, bringing in a pair of 230-kV lines. The other part, they focus very differently because of who their ownership is. That is one reason that I think the REC is more cautious in getting involved in renewables—their customers are the owners whereas Xcel is much more remote as far as influence one can have on them....it's like the big giant that nobody can talk to, whereas the REC has its ears open. The corporate world, when they have their mind made up to do something it is difficult to make changes and sometimes you have to go through some kind of embarrassment such as news media to bring things around a little bit.

This suggests a certain feeling of powerlessness to affect the outcomes of utility-scale energy decision-making.

11.2.6 POLICY AND REGULATORY DISADVANTAGES

One of the types of disadvantages of regional development of CSP that stakeholders mentioned in the six counties that compose the SLV involved disjointed county-by-county regulation. One Valley stakeholder said:

It's disjointed now. So you could have one county...Saguache has very lax building restrictions and it's a very poor county, so say if they want to attract solar into their county, that then impacts the surrounding counties because if their incentives are different than other counties...with six different counties in the Valley if you don't coordinate the plan, it is going to be disjointed.

The regulatory situation at the local level may need further development, a coordinated regional approach to be implemented. This could be expensive. Another point is that the BLM has to do a NEPA analysis, and that federal regulations have not caught up with CSP development. On the other hand, CSP development on private land requires no NEPA analysis. Finally, the point was made that the investment tax credit of 30% doesn't help developers because there is no way to turn it into money.

Although these disadvantages could be counted as barriers or as economic effects, the lack of adequate regulatory capacity would have negative consequences of various kinds for the people of the Valley. In the meantime, counties might compete with each other for siting of a CSP facility and each county may not be able to handle the regulatory aspects as effectively as might be optimal from the local perspective.

11.3 COMMUNITY SUPPORT AND OPPOSITION

11.3.1 INTRODUCTION

The stakeholders were queried on their views about whether the community supported or opposed the idea of CSP development. In addition, the interviewees were asked who the important stakeholders are whose views should be taken into account in CSP and transmission siting decisions. Finally, interviewees were asked to express their opinions about who favored CSP development and who, if any, opposed it.

The importance of understanding community support and opposition was underscored by one of the outside-Valley interviewees who said:

No one is really bird-dogging this issue, because—think about it from the developer's standpoint. They want to put in as much time as is needed to be successful in a competitive solicitation and so it is unlikely that they are altruistic enough to go on a San Luis Valley-wide educational campaign without having been the recipient of an award. Once that happens, I think that if the awards are strategic, they could do just that to educate the citizens on what their technology is, what they are intending to do, but the dynamic that has happened—it takes time and it takes money, and it is not necessarily needed in order to respond to an RFP. Until someone wins an RFP and proposes to develop a facility in the SLV, we won't know what the people think.

In this section of the report, the following materials are covered:

- A comprehensive list of stakeholder groups
- Self protection in the Valley
- Stakeholders' characterization of sentiment toward CSP development and transmission in the Valley.

Inevitably, interviewees mentioned the low levels of public knowledge about CSP, and also shared concerns about the self-protective nature of Valley residents, and the issues that are shaping opinion about CSP in the Valley.

11.3.2 STAKEHOLDERS IDENTIFIED RELATIVE TO CSP IN THE SLV

The interviewees were asked: What parties, organizations, or groups do you consider important stakeholders with respect to CSP development in the San Luis Valley? The responses to this question have been content-analyzed. At least 125 stakeholder groups (and a few key individuals) were identified, and these represent the most important stakeholders relative to the proposed project; however, they should be viewed only as a partial list. Although the stakeholders identified vary in the influence they might have on the CSP and transmission siting decisions, the length of the list illustrates the complexity of the web of socio-economic and environmental interests in the SLV.

Agriculturists
- Farmers in general (and their organizations)
- Potato growers
- Alligator farmer

- Ranchers (and their organizations)
- SLV Cattlemen's Association
- Colorado Farm Bureau
- Colorado Potato Administrative Committee
- 25 x '25

CSP developers/trade associations

- SkyFuels
- Abengoa Solar
- Ausra
- Interwest Energy Alliance
- Solar Energy Industries Association

Environmental organizations

- Audubon Society
- Clean Energy Action
- Colorado Natural Heritage and Rocky Mountain Bird Observatory
- Colorado Wild
- Colorado Wildlife Federation
- Crestone Sustainability Initiative
- Cultural Heritage Center
- Quiet Use Coalition
- Nature Conservancy
- National Wildlife Federation
- Playa Lakes Joint Ventures
- Rio Grande Headwaters Land Trust
- Sangre de Cristo National Heritage Group
- Sierra Club
- SLV Ecosystem Council
- SLV Citizen's Alliance
- Western Resource Advocates
- Wilderness Society
- Wildlife groups

Economic development

- Alamosa Convention and Visitors Bureau
- Bankers
- Chambers of Commerce
- Colorado Renewables Conservation
- Connecting Colorado Renewable Energy Resources
- SLV Resource Development Group
- Colorado Energy Forum (representing private investment companies)

- Optisolar
- Upper Rio Grande Economic Development Commission
- Valley Courier (Alamosa newspaper)

Educational institutions
- Adams State College
- Trinidad State Junior College

Elected officials
- Western Governors' Association
- Then-U.S. Senator Ken Salazar (now Secretary of the Interior)
- U.S. Representative John Salazar and his staff
- State Senator Gail Schwartz
- Governor Bill Ritter
- Boulder City Council, Ken Wilson (named as having an interest)
- Ex-State Senator Gigi Dennis

Federal Government Electricity Agencies
- Federal Energy Regulatory Commission (FERC)
- North American Electric Reliability Corporation (NERC)
- Western Electric Coordinating Council (WECC)

Government/Federal
- Baca National Wildlife Refuge
- Bureau of Land Management
- Environmental Protection Agency
- Federal Energy Regulatory Commission
- National Park Service
- National Renewable Energy Laboratory
- Sand Dunes National Park and Preserve
- U.S. Forest Service
- U.S. Department of Agriculture
- SLV Resource Conservation and Development initiative, U.S. Department of Agriculture

Government/State
- Governor's Energy Office (GEO)
- State legislators
- Clean Energy Development Authority
- Colorado State Land Board
- Colorado Division of Labor Workforce Center

- Colorado Department of Natural Resources
- Colorado Department of Transportation
- Colorado Department of the Treasury
- Colorado Public Utilities Commission
- Department of Water Resources
- Colorado Division of Wildlife

Government/NGOs/Local and Regional
- Alamosa City Manager
- Mayor of Crestone
- County commissioners (Alamosa, Conejos, Costilla, Mineral, Rio Grande, Saguache)
- Huerfano County Commissioners
- Alamosa City Manager
- SLV GIS Authority
- Historic Advisory Preservation Committee for Alamosa
- Rio Grande Water Conservancy District
- Upper Rio Grande Economic Development Commission

Landowners
- Crestone Property Owners Association
- Trinchera Ranch (Louis Bacon)
- Billy Joe "Red" McCombs

Law firms
- Energy Minerals Law Center
- Water law teams
- Front Range attorneys

Private companies
- A&J Solar
- Black and Veatch
- Blake Jones, Namaste Solar Electric (a PV entrepreneur)
- Black Hills, LLC
- Bill Clark Trucking
- Natural Power
- SunEdison

Service organizations
- Alamosa Wastewater Treatment Plant
- Hospital in Alamosa

The public
- Citizens
- The entire population
- Electricity customers
- Local residents
- People who live near the transmission lines

Solar and sustainable resource advocates
- Southwest Energy Efficiency Project (SWEEP)
- Clean Energy Action
- Colorado Solar Energy Industries Association (COSEIA)
- Colorado Renewable Energy Society (CRES)
- Solar Alliance
- SWAT (a Southwestern transmission organization)
- San Luis Valley Solar Association
- San Diegans for Smart Energy

Spiritual groups
- Shuma Institute

Utilities and transmission line owners
- San Luis Valley Power Authority
- San Luis Valley Rural Electric Cooperative
- Tri-State Electric Generation and Transmission Association
- Xcel Energy
- Public Service Company
- Colorado Rural Electric Association
- Colorado Independent Energy Association

Water interests
- Rio Grande Water Conservancy District
- San Luis Valley Water Group
- Water law teams of Front Range legal firms
- SLV Water Protection Coalition
- Water Watch Alliance

11.3.3 SELF PROTECTION IN THE VALLEY

One theme that emerged from the interviews was self-protection in the Valley. Seven of the 15 Valley respondents talked about the strong community aversion to outsiders coming into the Valley and taking advantage of its residents in any way. Even the interviewing for this study was seen by a few Valley stakeholders as an intrusion of outsiders into local affairs. One outside stakeholder with connections in the Valley put it this way:

Most rural communities have fear; they're skeptical of folks from outside coming in and doing stuff. It's an interesting mix of emotions and you don't want someone from the outside coming in, but you don't believe you have the capacity to do it yourself. That's true for many rural communities—it's a rural psychological mindset.

A Valley stakeholder put it succinctly:

The biggest issue on the table is will we do it ourselves or will it be done to us and for us?

Other respondents from the Valley said the following on this point.

We were looking at concentrated solar and what they were able to produce now in terms of megawatts and, obviously, a lot of this information came out of California, and then how much land base that was used and then multiplied that land base to come up with 5.6 gigawatts. So we started looking at that and thought—my God, that's an industrialization of the Valley floor. And of course they need water, concentrated solar needs a fairly substantial amount of water, so the valley has the three elements they are looking for which is one of the reasons that European companies are starting to court the Colorado State legislature. And that's happening now. It has been happening.

Our people are very vocal...if they came in and worked with us, there wouldn't be much opposition. But if an outside group tried to come in and strong-arm us, this is a tough community. We want to protect the environment and to protect against big government and corporations. If it is done properly and through the right channels...so, really, it is how it is done.

The reference to California is an important indicator that local residents opposed to a proposed transmission line to export renewable electricity from the SLV were in communication with an organized group opposing transmission lines for renewable electricity in California (San Diegans for Smart Energy Solutions). Local area organized opposition groups can and do share information and resources with each other in other parts of the country. Interestingly, the issues posted on the San Diego group's web site track closely with some of the claimed advantages of the proposed transmission line into the SLV, including the following.

- Serious "power gap" without the transmission line
- Proposed line is needed to transport renewable electricity to meet RPS goals
- Protects against fire hazard
- Cost-effective and beneficial to ratepayers
- No significant environmental impacts
- Ensures energy reliability
- Good for the local economy and fosters job creation.

The opponents to transmission attempted to counter each of those points.

This preliminary evidence suggests that these themes will likely recur in areas of the Southwest wherever transmission lines for renewable electricity are proposed. Given the commitments of the Western Governors' Association and other key organizations to developing transmission to bring renewable electricity to cities, understanding the dynamics of these opposition groups is evolving as an important national issue.

11.3.4 COMMUNITY POSITIONS FOR AND AGAINST RENEWABLE ENERGY IN GENERAL

Stakeholders were asked: "Thinking about the SLV community as a whole, and also about different groups within the community, what would you estimate is the level of support for renewable energy development in general—positive, negative, or don't have an opinion?"

None of the stakeholders within or outside the Valley said that the community was opposed to the development of renewable energy in general.

In fact, the general sense from the interviews was that groups in the Valley strongly support renewable energy. An illustrative quote is as follows.

In the Valley, I would say [support for renewable energy] is extremely high. I would say a supermajority, how's that? More than a majority.

Tremendous, very high—without exception folks are buying into renewable energy development.

11.3.5 COMMUNITY POSITIONS FOR AND AGAINST CSP DEVELOPMENT

Stakeholders were also asked: "How about levels of support for the proposed CSP facility?" Responses differed from the broad support mentioned for renewable energy generally. Of the 15 Valley stakeholders, only two said that the community favors CSP development, and one of those said the community didn't understand what CSP is. Support for solar development is based largely on the favorability toward the SunEdison PV plant near Mosca.

Support for Sun Edison was high, so why would people not jump on board and be supportive of a CSP project?

A Valley stakeholder said, "The reason the Tri-state people won't build PV systems is the cost ($0.22/kWh)." Speaking about one of the citizens at a public meeting, this stakeholder said:

[The opponent] thinks we can run the entire Valley on solar and not have any need for transmission and most of these people have the idea that we could become isolationists and serve our own needs. Well, that's taking some steps back in time that nobody I'm familiar with is willing to take.

On the other hand, half of the stakeholders outside the Valley (5 of 10) said they thought the community was favorable to CSP development.

I would say [the level of support] is high. Just from what I've heard from economic developers, from citizens, from residents. I can't say I've talked to a lot and, of course, most who have talked to me talked me up, so it's not...I wouldn't say I have rubbed shoulders with a representative sampling of the valley residents, but from what I do hear and read the support is quite high.

Another outside stakeholder said:

I would say mostly positive. Again, I am basing this [on my experience]. I take it that their elected representatives were in support. [At] a town meeting in Alamosa last year [there were] about 65 people...and I would say that except for the reservations....Well, there were two things. One were some reservations about transmission lines, although the crowd, on average, overall decided it would probably be worth it, and [there were] a lot of advocates there for more decentralized solar instead of just centralized stations.

A third said that the people are positive "but not terribly knowledgeable." Nearly half (n=7) of the Valley interviewees said they believe the SLV community was mixed in its reaction to CSP development. The situation is complex. It was said that the county commissioners need help in understanding capacity. It was also noted that Representative Salazar's aides were hearing from opponents. Some quotes from stakeholders inside the Valley exhibiting the community's mixed reaction are as follows:

I think there are certainly cultural issues and issues of sustainability. Commodifying of natural resources seems to be a big concern—the commodification of solar power.

Interviewer: *Does that translate to opposition to exporting power out of the Valley?*

Stakeholder: *Yeah, I think people are concerned about becoming an exporter of power. Again, the Valley wants to receive something in return.*
Interviewer: *And what does 'cultural issues' mean?*

Stakeholder: *Philosophy that solar energy and wind power are appropriate for farm and residential areas and sustainability, but on the larger scale it becomes more utility, becomes industrial. Even the idea of a major corporation being in control of a local resource and making money from it [is a problem]. There are some that want community control of it, some who don't care, and some who are just concerned about the scale of the development.*

Another Valley interviewee expressed it this way:

Interviewer: *Would the SLV community oppose a project that large?*

Stakeholder: *It depends on where it is located, but finding that amount of space would be very difficult. There is a large environmental contingency that is very active and would have concerns.*

Interviewer: *What types of concerns do you think they would have?*

Stakeholder: *So far they are concerned about it being an unsustainable renewable energy development, too large of a scale-- large plots of land for basically one industrial purpose. They have concerns about the water as well—the quantity of and types of solutions or lubricants used. Some people are overwhelmed by the size and taking agricultural land out of production and decline of a way of life.*

Interviewer: *So a concern about the cultural heritage of the Valley?*

Stakeholder: *Absolutely, absolutely. And, if they end up putting a large facility on the side of a mountain and a big shiny thing, that would raise a lot of concern.*

Interviewer: *So aesthetic concerns?*

Stakeholder: *Absolutely, the aesthetics.*

Interviewer: *What about locating a facility on retired agriculture land? How would that be received?*

Stakeholder: *I think farmers in particular would be open to that. It's just a matter of logistics.*

Another Valley stakeholder spoke of mixed reactions:

Interviewer: *So would you say in general people are favorable toward CSP or not?*

Stakeholder: *I think there are a lot of exaggerations, a lot of anxiety, and there is a lot of misinformation...the question is, one, are we going to see generation that is simply going to be shipped out of this valley with no revenues/resources left behind...what they're worried about is can there be a way to share in the profits/benefits of the industry, long-term jobs, not just short-term jobs, and is there a way to also have revenues that will be generated and stay in the region? I have some ideas about that.... And there's a push to have more, what they call, distributed generation, to put solar panels on all the houses and the schools and the buildings and, therefore, maybe don't use utility scale. You hear that push back about we don't want to change the culture.... I think you're in a place that could go one way or the other. You could have some development here...here isn't enough...I think the utilities* *are certainly going to control it because of the power purchase agreements, but the question is you want to have it integrate into the value and the culture of the region.*

Another Valley stakeholder said about the opponents:

They want it to just be for us only and not transmit anything out. They all want clean energy but they don't want to give up...have any environmental impact.

Outside the Valley the view of community support and opposition is different, as would be expected. An outside interviewee said:

I imagine it is like anywhere else. There're going to be people who are for it and people against it.

Another outside stakeholder said:

... that the citizens in the valley are highly interested and whenever there has been an event, an educational event, the rooms have been packed. They certainly understand the gravity of the situation and the need for energy and the need to educate themselves. I think they are understandably tepid about what the impacts might be, specifically to their local communities, for the reasons that I mentioned about a not well-understood technology and wanting to make sure that their local communities benefit from any development in the San Luis Valley.

Interviewer: *So it is clearly understood by the people but not by the.... You mean by the people themselves.... You don't mean by the technologists?*

Stakeholder: *I think it is a new technology to most people and I think there is a good amount of education or a lot of education that would probably need to happen in the valley and I think we need to turn that responsibility back over to potential developers* *to educate the citizens in the valley on what the technology is, what it isn't, to debunk some myths, but probably, most importantly, not to overpromise. I think...the feeling in the valley is a strong interest.*

Another stakeholder outside the Valley said:

From the different kinds of groups, I'd say anybody in the economic development area would probably be supportive with some caveats probably, but...

Interviewer: *What kind of caveats?*

Stakeholder: *I think that what I hear is probably a realistic concern, is that when there is a large plant to be built, of any kind, you can bring in labor, and can tend to come into an area, overwhelm the area (maybe, maybe not) and then leave the area. So, there's a kind of preference for local labor and that doesn't always happen. That was the case in Nevada for Nevada solar power, so that would be one of those caveats.... I think you might find groups that are leery of companies coming in and taking advantage of the area."*

11.3.6 DON'T KNOW WHAT STAKEHOLDERS FEEL

Three of the interviewees (two outside and one Valley) said that they didn't know what community response toward CSP development was. An outside Valley stakeholder said:

I don't know because I haven't wandered around there and asked people. At [other sites] public support has been tremendous, very positive. Economic growth, tourism, clean energy, it has been exciting. I think the same will be true in the Valley. Whoever develops the plant has to start meeting early with the local people. It takes a long time to explain to people—[questions arise], such as will the mirrors blind pilots? [One developer] had to fly over [a] plant and take a video showing that at no angle would the sun's rays be reflected up to the airplane.

11.3.7 PERCEIVED NEED FOR EDUCATION

Several interviewees stressed the need for public education. For example, one of the outside stakeholders said:

There might be a misperception by the public that these renewables are an answer. I don't think they understand the impact that, especially solar, has on the land. And that is you know—one square mile of solar facility equals so many megawatts and when you really look at that it's a square mile of nothing but a solar facility. There's not too much else that happens on that land.

Two Valley stakeholders said:

They don't know enough. If people were honest with you, I think they would tell you that. They just don't know because of lack of education. Adams State College could play a role with that if they wanted to—could be a place where educational forums could take place.

...there is no group that is standing out there saying we need concentrated solar power. There is no group that even probably knows what it is.

Pointing to the need for public education, another interviewee from the Valley stated:

[There was] a public meeting last night.... Some of those speakers from Crestone there...had no desire for any kind of transmission project and had the desire that we produce all of our energy from our own source.... [There have been] more and more meetings.

*Last week [there were] two meetings on transmission lines and all these issues get addressed every time [there is] a meeting....
[there will be] another series of meetings basically on the transmission line, [I think there should be] a focus group of people to deal with energy issues because there are huge issues today.*

11.4 TRANSMISSION ISSUES RAISED BY STAKEHOLDERS

11.4.1 INTRODUCTION

The interviewees were queried on their views of the most important considerations in the siting of a transmission line for renewable electricity. They were asked: "From the perspective of your organization, what are the most important considerations involved in power transmission?" Responses ranged from full support for the transmission through neutral, factual responses, to mention of community conflict about transmission, to opposition.

Stakeholders in the Valley said that Rio Grande, Alamosa, and Saguache Counties are interested in CSP development. It was generally agreed by knowledgeable stakeholders that new transmission lines would be needed if a CSP facility at a scale of about 200 MW were to be developed. One said: "The current Climate Action Plan models a 200-MW concentrated solar power facility somewhere in Colorado." Viewsheds, availability of water, and availability of transmission are the biggest considerations for siting CSP. This section discusses the issues raised in the interviews relative to transmission on a continuum from support to opposition.

11.4.2 SUPPORT FOR TRANSMISSION

As reasons to support transmission, Valley interviewees perceived the following: (1) the desire to export power as a way to bring income into the community, (2) the need for security and redundancy of the Valley's power supply, and (3) the need for improved power infrastructure.

11.4.2.1 DESIRE TO EXPORT POWER

We located 20 comments from stakeholders inside the Valley that addressed the need for transmission to export power. This is one of the most commented-on issues in the interviews. One of the Valley interviewees said:

> *Well, in some ways we have the ability to meet our needs with solar and provide for other people's needs, so we have done a little bit of research, even with people that live in places like Boulder and asked them if I have a megawatt of power and I'm generating in the San Luis Valley, would you have an interest in it? And we've got responses that include things like not only would I have an interest in it, if it is really renewable and can really guarantee that this project could be sited in the San Luis Valley, I will pay you a premium. I will give you a $1.30 for your $1 worth of power and I will buy it by subscription, I will pay you a year in advance so you have the money to help build the facility.*

> *Yeah, it's about 200 MW as a rule of thumb for a number to use. So if you went to 200 MW, doubled its size, you would meet the demand and then you would have the ability to expand to 400 megawatts and instead of importing 200 you'd be exporting the extra 200 so now your renewable energy generator is exporting 200 MW and generating the same income that used to leave here as a payment to a utility.*

> **Interviewer:** *Okay, so the idea is to own [one or more CSP facilities] locally and first meet local demand.*

> **Stakeholder:** *That is going to be the most palatable environmentally.*

> **Interviewer:** *Okay. What about transmission? I mean transmission is a big issue or apparently a big issue, so what about that?*

Stakeholder: *Well, 100 MW, should we put this on the ground, we meet all the conditions that you and I described to date, then we would be meeting half the demand of the San Luis Valley.... And second export and bring money into the Valley by selling power.*

An outside stakeholder described it this way:

Really no one can build anything in the SLV until there's a transmission highway to get it out.

11.4.2.2 NEED FOR SECURITY AND REDUNDANCY

Valley stakeholders talked about their concerns about the vulnerability of the Valley's electricity supply. Ten comments fell into this category. One interviewee from the Valley said:

Interviewer: *You mentioned transmission as a limiting factor?*

Stakeholder*: Yeah. But there is a proposal for Xcel and the REC to go together and put a pair of 230-kV [transmission] lines over La Veta Pass from the east, which would be a good idea because of the vulnerability that we are in right now because it all comes over from the north.*

Another Valley stakeholder said:

It comes from the north, two transmission lines, and last night we even left the Homeland Security comment on the table that two guys with one chainsaw could take all the power out in this valley. There is no redundant power. All the poles....if you cut four telephone poles you can shut down the power in the entire San Luis Valley. It is completely inappropriate for any future development. There is a huge significant issue about redundancy in transmission lines.

Another stakeholder was asked if they had any thoughts or feelings about the energy leaving the SLV to help Denver or the Front Range. The response was:

Yes, positive as long as it promotes revenue. But concerns about transmission—what will it take to handle the transmission of solar power. We don't want to export the water. If there is a win-win then the Valley would be all over it.

Another Valley stakeholder said:

So the Valley basically right now uses about 155 MW and I believe you need 1,000 MW to equal one gigawatt, so then the question becomes, geez, what is all that other potential for? Well part of it is redundancy so the Valley has another way to get energy if something happens; the other is for the Valley to grow and possibly invite industries that have never been able to locate here before. That's what's being promoted. And then the other piece is that it is a valuable potential to be able to export power. For the first time the Valley will become an energy producer, and obviously one of those ways the Valley will be able to produce energy is through concentrated solar.

11.4.2.3 NEED FOR IMPROVED POWER INFRASTRUCTURE

At least one Valley respondent said that the SLV needs more power.

Interviewer: *So they must be thinking that there is going to be renewable energy development in the Valley if they're building this line.*

Stakeholder: *We needed the power in as much as eight or nine years ago, so the initial thrust for that line was us needing power to support our developments, our growth in the western area,*

South Fork, Creede. When renewable energy came around and NREL named us a spot for really good solar and they passed the renewable energy bill, all of those things said to Xcel we're going to need transmission from the San Luis Valley.... [It will happen] one way or another. And it should be.... we don't put any money in, it is just for our system, and Xcel and Tri-State pay for it. It is about an $80 million project. It should be completed by 2012 or 2013.

Interviewer: *And why are they doing this?*

Stakeholder: *Because we are running out of power in the valley. The last transmission line built in San Luis Valley was 35 years ago.*

Although existing transmission lines could probably support an additional 50-100 MW of CSP development, Valley stakeholders made several comments suggesting that the size of the CSP plant needed to be as large as 200 MW. As noted earlier, this size of facility would allow local demand to be met first and would begin to yield additional electricity for export out of the valley.

Another local stakeholder talked about the scale of a possible CSP facility.

Is this the one [a Valley expert] keeps talking about that is 5,000 acres and enough power to power six planets? ...We talk about it all the time because it is considered very important for our economic development. If you take a resource like that and don't leverage it, then you're really missing the boat. We are "green" here in the Valley and it resonates with us. Our agriculture industry is having problems because of water and depletion of the aquifer. And since agriculture is our number one industry, we need to diversify. Tourism is our number two industry, which only represents 11% of our economy. We have 80,000 square miles here in agriculture production, so...if some of these farms converted to a product that did not use so much water that is a good thing.

An outside stakeholder said:

> *What sets solar thermal apart from PV? PV uses semiconduc-*
> *tor—electrons —generates electricity directly. Solar thermal is a*
> *large-scale parabolic trough—above 5 MW it's cheaper than PV.*
> *Below 5 MW, PV is cheaper. So solar thermal gives you the scale.*
> *200 MW is the sweet spot. CSP has scalability. You would never*
> *have a 200 MW PV system.... The characteristics of thermal are*
> *that it has more thermal inertia—in the tube is oil at 750 degrees.*
> *If a cloud goes over it, it still stays hot, so it has more even per-*
> *formance. PV production is spiky because when it is cloudy, the*
> *output drops dramatically. So each time a cloud passes over a PV*
> *system, the production goes way down. Thus, the grid has to be*
> *able to handle spikes. There is no such limitation for solar ther-*
> *mal, it's very predictable and even....*

Stakeholders said that, in the Valley, the transmission issue was as big as the water issue. Yet, it seemed to be fairly generally agreed that new transmission had to be built if a CSP plant of any economic size (e.g., 200 to 300 MW) is to be sited in the Valley. Looked at another way, a CSP plant up to 200 MW could be accommodated on existing lines, but anything beyond that could not.

11.4.2.4 NEUTRAL COMMENTS

Two quotes below illustrate a neutral, fact-based explanation of the transmission situation without exhibiting a position.

> *It's been all over the newspapers that the SkyFuel wants to do*
> *1 gigawatt power—they won't do it all at once though because*
> *we don't have the transmission to handle that. They are going*
> *to have to build out if that ever happens. But there again, that's*
> *5,000 acres.*

And,

Number one is current availability of transmission, and we are currently maxed out. There are three coming over Poncha Pass— small, medium and larger. There is one proposed to come over La Veta Pass.... Second concern is size of a new transmission line— can it meet current demand and possible new development. I think solar development will go beyond the capacity of the new line (230 KV) that Tri-State and Xcel are proposing. If at some point we are going to develop solar power and transport it out of the Valley, we need to address this issue. But how do we finance this?

11.4.3 OPPOSITION TO TRANSMISSION

As reasons to oppose transmission, Valley interviewees mentioned (1) fear of industrialization of the Valley floor, (2) aesthetic concerns, (3) feared loss of control, (4) wildlife concerns, and (5) legal-regulatory concerns.

11.4.3.1 FEAR OF INDUSTRIALIZATION OF THE VALLEY FLOOR

A few comments appeared to oppose transmission lines. Fear of un-controlled industrialization of the Valley floor was expressed in the following quote:

I don't think anybody in the valley has a problem with a one giga-watt transmission line. The problem is once that transmission line gets in here, what are the possibilities of taking advantage of that because the loop will then all be there, the loop will be formed, and then other industries will be very attracted to the fact that it is in now and they may want to develop and build on that and they may want to take it to 5.6 gigawatts.... All these companies aren't going to invest in the valley for 1 gigawatt. They're not. They're going to want more and that's the quandary. And unfortunately that's what concentrated solar represents.... So are we really thinking in an antiquated way by thinking in terms of concentrated solar? Will we end up coming up with much more efficient ways of being able to distribute solar other than by concentrating it?

Also, concerns about public health and electro-magnetic fields (EMFs) were said to be concerns of neighbors living near the proposed lines.

11.4.3.2 AESTHETIC CONCERNS

A few comments decried a proposed transmission line because it would be "on our vistas" and would be aesthetically displeasing, "going over the mountains."

It was noted that one of the Valley's counties opposed transmission lines.

Costilla is most concerned with transmission lines being placed across their county—aesthetic concerns, and also they want to be sure that they receive local benefits. In particular, their largest landowner, Mr. Bacon is concerned about transmission lines going across his private land, which is used for ranching and wildlife. He might have concerns about future property development.

11.4.3.3 FEAR OF LOSS OF CONTROL

As noted earlier, Xcel Energy and Tri-State would be granted eminent domain should they be granted the CPCN for the transmission line. Some comments expressed fear of loss of local control of property legally condemned under eminent domain.

11.4.3.4 WILDLIFE CONCERNS

Concerns were expressed about the possibility that the transmission line would bisect wildlife corridors in the La Veta Pass and Trinchera Ranch areas and that transmission lines can be hazardous to birds. According to some residents, power poles can give raptors an advantage over the other species because transmission lines are required to provide raptor perches.

11.4.3.5 LEGAL-REGULATORY CONCERNS

As described in Part Two of the report on the transmission controversy, Louis Bacon, who owns the Trinchera Ranch, has hired attorneys to oppose Xcel Energy and Tri-State's application for a CPCN to build the transmission line. In interviews, Valley stakeholders had identified Louis Bacon's opposition to a proposed transmission line.

End of excerpt. To view the full report, please use the citation information on the first page of this chapter.

CHAPTER 12

THE PROMOTION OF DOMESTIC GRID-CONNECTED PHOTOVOLTAIC ELECTRICITY PRODUCTION THROUGH SOCIAL LEARNING

GREG HAMPTON AND SIMON ECKERMANN

12.1 BACKGROUND

12.1.1 PUBLIC UNDERSTANDING AND POLICY CONTEXT OF GRID-CONNECTED PHOTOVOLTAIC ELECTRICITY PRODUCTION IN AUSTRALIA

Use of grid-connected photovoltaics (GCPV), which involve the installation of photovoltaic panels (PV) on a roof or external wall and generating electricity for use and/or export to the grid, has been in its infancy in Australia. GCPV is a renewable source of power without environmental cost in producing electricity once installed or the need for land to be used and has minimal transmission or distribution cost [1], while having up front panel, inverter and instillation costs. Much of Australia has an ideal climate for generating solar energy with these technologies, as well as wind energy, and policy support by Governments in Australia for use of renewable energy has emerged over the last 5 to 10 years in large part in response to international targets for greenhouse gas emissions. Production of electricity in Australia has historically been predominantly from burning coal, and Australia retains a very high per capita emission of greenhouse gases. Government policy to increase use of renewable solar energy generation has been enacted with economic instruments such as rebates and gross feed in tariffs by Federal and State levels of government (electricity grids between states are not fully interconnected nationally). However, there has been meagre research on public familiarity and attitudes towards GCPV and the associated subsidies and feed in tariffs.

This article considers processes of social learning about the economic instruments and decision making, given technology and installation options for photovoltaic electricity production. Empirically, we report on social learning with randomly selected public participants from a regional area south of Sydney, New South Wales in deliberative workshops undertaken at the University of Wollongong in 2005 and 2012. These workshops provided participants with a general introduction to the practical and financial feasibility of domestic production of electricity through installing photovoltaic panels with inverters and the economic instruments involved in residential installation.

The enhancement of social learning about residential photovoltaic installation is an important component of effective policy implementation for renewable energy in the Australian context. Meeting renewable energy targets is in part dependent on citizen take up of renewable energy systems, while it has been left to citizens to initiate the implementation of this technology through a renewable energy certificate system.

Large-scale Generation Certificates are currently provided by the Australian Federal government for the installation of capital equipment in large schemes [2], and as the name suggests, this is likely to be implemented by large-scale generators. Under a parallel small-scale renewable energy scheme, Small-scale Technology Certificates (STC) are provided for domestic installations [3], and it is in this area that citizen social learning is of importance.

This paper considers how social learning processes could be used by such citizenry to aid information flows, decision making and policy implementation. Having set the Australian context for decision making by citizenry in this introduction, the next section considers the decision to invest in PV technology. The concept of social learning in relation to decision making for small-scale PV investment is then reviewed in relation to technology characteristics (PV system installed costs, expected performance, aesthetics, building integration) and government policy instruments (subsidies, tariffs and rebates) and their interaction in informing expected return on investment, before describing methods for and reporting on attitudes to PV decision making emerging from deliberative social learning workshops run in 2005 and 2012. Finally, the results from the workshops in relation to understanding and attitudes reported are discussed in relation to the differences in policy context and framing of PV investment decisions in 2005 and 2012, drawing out lessons for future policy and use of social learning methods.

12.1.2 CITIZENRY DECISIONS TO INVEST IN PV UNDER ALTERNATE POLICIES

A decision to invest in installing a small-scale photovoltaic installation depends upon the inclinations of residents and their level of knowledge of photovoltaic technology, particularly the relative environmental and economic costs and benefits which may accrue from such installation. An STC can be transferred to and redeemed by a photovoltaic installer, reducing the capital investment costs of a typical 1.5-kilowatt (kW) installation to around Australia $3,000 at the time of the workshops in early 2012.

This system of STCs was preceded by a Federal government rebate scheme between 2000 and 2009 which resulted in the installation of 107,572 units generating 128 megawatt (MW) [4]. The previous program has been regarded as being ineffective environmentally and inefficient in terms of failing to promote understanding and acceptance of residential photovoltaic installation. The suggestion is it would have been less costly to employ other strategies such as standard social marketing [4]. Further, such alternative strategies could have mitigated negative publicity and associated framing of investment decisions for residential photovoltaic installation when prices for STCs or State arrangements for feed in tariffs for generated electricity changed in 2010. This, consequently, could also have mitigated the uncertainty and boom and bust cycles for associated manufacturing and installation industries.

Wustenhagen et al. [5] reviewed research on the social acceptance of wind energy and pose some questions which might also arise for citizens who install photovoltaic systems in their residences. Much of the research which has been conducted on social acceptance and renewable energy has been related to wind energy, nuclear power and geothermal, and has not focussed on photovoltaic panels on residences. For example, social acceptance of wind energy has shown the type of landscape to be a critical factor [6]. Analogously, the siting of PV solar panels may be an important issue, particularly if large-scale photovoltaic installations become more plentiful. There is some evidence that the visual intrusion of residential panels is associated with a decision not to install panels [7]. Australians may have concerns about the visual amenity of having panels installed on a residence, in relation to which the workshops developed for the current project assessed perceptions of traditional panels versus building-integrated panels.

There are likely to be varying levels of community engagement and education to promote the installation of a residential photovoltaic system, depending on factors such as whether or not citizens were residents in one of the designated seven Solar Cities in Australia [8]. Education of citizens in general about photovoltaic technology in Australia consists currently primarily of static website information. There have been no apparent government initiatives for citizens to pursue this information or highlight the expected financial returns from investing in and using electricity generated by residential photovoltaic systems.

Assessing citizen attitudes towards and motivation to investing in PV systems in Australia needs to distinguish between wholesale price levels for exported feed in tariffs, effective electricity cost offsets with direct generated use and government subsidized tariffs. Such subsidies can provide incentives to install domestic photovoltaic electricity generation technology but in turn need to consider the real expected return and payback periods given the expected performance, and up front capital costs, of systems.

In addition to Commonwealth Government cost offsets accruing from STCs, State governments across Australia have provided a range of domestic electricity offset and feed in tariff arrangements and associated incentives. These initially ranged from State tariffs with typical small-scale (less than 10-kW systems) systems for net export of electricity in excess of domestic use into the grid (South Australia (SA), Victoria (Vic), Western Australia (WA), Queensland (QLD)) of amounts varying from Australian $0.44 to Australian $0.60/kWh over 10 to 20 years, to gross feed in tariffs in New South Wales (NSW) and the Australian Capital Territory (ACT) over 6 to 20 years from Australian $.040 up to Australian $0.60/kWh [9]. In NSW, where our study is based, an Australian $0.60 gross feed in tariff applied to installation and supply agreements in NSW signed from 9 November 2009 up to midnight 27 October 2010 for electricity generated by PV panels up to 31 December 2016 under the Solar Bonus Scheme program. This program essentially covered the capital investment costs by households of a typical 1.5-kW system installation (around Australian $6,000 during 2010) by the end of 2016, given that under this tariff such a system reduces energy costs or generates income of approximately Australian $1,100/year. This NSW program has more recently been framed as burdensome by the current NSW State Government, but the program did ensure a rapid uptake of photovoltaic technology in the State, while not committing tariff funding beyond 2016 unlike other state schemes. For agreements signed after the 28 October 2010, the gross feed in tariff in NSW was reduced from Australian $0.60/kWh generated to Australian $0.20/kWh, but a newly elected state government in April 2011 effectively removed any government subsidy, moving to a tariff to reflect the cost of undistributed wholesale electricity. For PV panel installation and supply agreements in New South Wales at the time of the workshops conducted in 2012, residents could receive around an average Australian $0.07/kWh for electricity exported to the grid—a price which is based on the spot price which a wholesaler would receive for electricity generated in the state [10]. Alternatively, if the residents in NSW were to utilize electricity generated by their photovoltaic installation in their own home, this would be financially equivalent to what they would pay for utilizing electricity from the grid, typically Australian $0.22 to Australian $0.24/kWh, unless use is off-peak. Hence, offsetting domestic use in peak periods when PV electricity is generated would, on average, provide three times the price which they are offered for exporting electricity through a net feed in tariff. However, to utilize PV energy generated at their home, residents would have to utilize this electricity at the time it is generated (during sunlight hours) which may not be feasible for many residents given their patterns of energy consumption.

Extensive publicity and political attention around the reduction in gross feed in tariff rate to Australian $0.20/kWh for systems signed off after the 27 October 2010, and further falls in the exporting rate when investing beyond 28 April 2011, has negatively framed public perception of investing in domestic photovoltaic technology in NSW after 27 October 2010. Indeed, beyond 28 April 2011, while the Independent Pricing and Regulatory Tribunal (IPART) recommended a non-mandatory price per kWh around the wholesale average undistributed price of Australian $0.052 to Australian $0.103 per kWh in 2011 to 2012 and Australian $0.077 to Australian $0.129 in 2012 to 2013 [11,12], electricity retailers effectively paid at the bottom of these ranges, if anything, for PV exported electricity. Hence, there was marginal if any return received from investing in a gross metered system, depending on retailer after 28 April 2011, with net meters clearly preferable when offsetting domestic electricity use with PV energy generated during daylight hours (effectively Australian $0.20 to Australian $0.28, depending on provider over 2011 to 2013).

Importantly, the rate of payment for PV energy generation with investment after 27 October 2010 was publicly framed as a loss relative to that previously received at Australian $0.60c/kWh. Public framing and focus on such perceived losses is, following prospect theory and loss aversion with higher valuing of perceived losses than gains [13], expected to lead to sharp falls in GCPV investment and installations. However, continual significant decreases in capital costs faced by households for installing systems after 27 October 2010 and increasing prices for grid electricity mean that while the payback period for installing domestic solar panels may have increased somewhat, the long-term rate of return to households on a typical 1.5-kW system has in fact increased over time. For example, a reduction for the installed cost to a household of a 1.5-kW system to

$3.000 with an agreement signed in early 2011, with an Australian $0.20c feed in tariff until the end of 2016, results in an expected 7- to 9-year payback period. A 1- to 3-year payback period beyond 2016 depending on the extent to which feed in tariffs are higher than Australian $0.30 beyond 2016, with converting from gross to net meters and domestic generation offsets daylight consumption and doubles the return on initial capital investment relative to an Australian $6,000 system beyond that. Hence, falls in capital costs after October 27 2010 result in a higher rate of expected overall investment return with an outlook beyond 10 to 15 years despite the lower rate of tariff to 31 December 2016 in NSW. Indeed, the long-term return continues to rise with continuing falls in capital costs of installation, even with the further fall in effective gross tariffs paid to the end of 2016. Currently, 1.5-kW systems in NSW installed with net meters offset energy use during daylight (about Australian $0.25 to Australian $0.30/kWh) and receiving what amounts to an average wholesale price (around Australian $0.077/kWh) with an appropriate choice of retailer can be installed for as little as Australian $1,500. This results in a 4- to 12-year payback period, depending on energy use offset and the extent to which the retail and wholesale price for electricity can be expected to increase to 2016 and beyond. If, as reported by IPART, on average, two thirds of PV generation is consumed and one third exported to the grid, then the average payback period with a net meter and 1.5-kW system in 2013 would be approximately 6 to 7 years, depending on the rate of increase in electricity prices [12]. Those consuming all PV energy generated would have a payback period of approximately 4 years, those who export 50% about 9 years, and those exporting all energy generated around 12 years.

Beyond 2016, the net-metered PV systems with Australian $1,500 capital installation costs in 2013 will have twice the return on capital of fixed systems installed for Australian $3,000 and four times the return on a system installed for Australian $6,000 in 2010. Those with a gross feed in meter who received Australian $0.60 and Australian $0.20 will also face costs of converting to a net meter post 2016 to allow for the higher effective returns from offsetting electricity use during PV generating hours.

Consequently, the processes of social learning and the investigation of public perceptions is becoming increasingly pertinent to effective promotion, informed investment decisions and efficient implementation of such domestic PV systems in NSW. Further, while the above consideration of returns on investment is framed in the context of subsidized gross-metered tariffs in NSW, similar calculations and arguments for approaches to informed promotion with social learning apply in other jurisdictions (Australian States and elsewhere) where reduced capital costs and increasing energy prices result in increased long-term return on investment for net-metered PV systems despite removal of previously subsidized tariffs for fixed periods. However, despite the general and increasing need to have informed public perception and investment decisions, the promotion of domestic utilization of photovoltaic technology in Australia remains very limited. This is primarily limited to Internet websites created by government departments and installers of photovoltaic panels and occasional printed postal information provided by utilities and retailers.

It is suggested that in order for citizens to be more knowledgeable about the environmental and financial returns from installing photovoltaic technology, consideration should be given to the utilization of processes of social learning within various modes of educating citizens. These processes of social learning can be incorporated in community information sessions and other forms of media promotion.

12.1.3 SOCIAL LEARNING

Social learning is briefly reviewed to articulate essential components required for a social learning process in relation to domestic grid-connected photovoltaic systems. The facilitation of social learning is critical to the process of developing citizen proficiency in understanding science and technology. It is designed 'to enlarge the citizen client's abilities to pose the problems and questions that interest and concern them and to help connect them to the kinds of information and resources needed to help them find answers' [14]. Schusler et al. ([15], p. 311) defined social learning 'as learning that occurs when people engage one another, sharing diverse perspectives and experiences to develop a common framework of understanding and basis for joint action'. Keen et al. [16] propose a model of social learning comprised of reflection, systems orientation, integration, negotiation and participation. They include the economic system in their system orientation which is relevant to financial aspects of the installation of grid-connected photovoltaic production of electricity in domestic premises.

In social learning, the social condition can be changed or altered, especially changes in how one perceives their personal interests compared to and connected with the shared interests of their community [17]. Conditions for social learning need to be conducive for meaningful dialogue and interaction to occur between experts and non-experts that results in an environment for thinking and learning together. Some aspects of a program that can promote social learning include providing an atmosphere of open dialogue and transparency of information, opportunities for repeated meetings and gatherings, access to expert support, face-to-face small group work, site visits and tours, unrestricted opportunities to influence the program process and political support for the process [17]. In this project, we developed a deliberative workshop format which provided participants with information on photovoltaic technology, building integration of photovoltaic technology, government economic instruments designed to promote domestic use of this technology and the opportunity to reflect on this information.

The sharing of diverse participant perspectives involves providing a public with an opportunity to discuss their experience and local knowledge of an issue [14]. Local knowledge is often tacit knowledge which can be rendered explicit through deliberation [18,19]. In the deliberative workshops, we built upon participants' local knowledge by referring to other forms of renewable energy, already in existence in the local area, which they were familiar with and provided participants with the opportunity to elaborate on the issues which arose for them in the workshops.

Deliberative workshops incorporating social learning provide the opportunity for participants to consider technological and domestic issues in depth as well as elicit issues not considered by researchers. The workshops provided participants with an opportunity to share their views on energy production and individual usage and the personal, economic and social conditions which influence the uptake of photovoltaic systems in domestic dwellings. The participants were provided with opportunities to discuss similar technologies, such as wind energy and a local trial of ocean tidal generation of electricity, which informed their understanding of similar contexts for photovoltaic electricity production. In order to develop the workshop process, social science research relevant to citizen attitudes towards and knowledge of photovoltaic technology was reviewed. This guided the choice of relevant content for the development of the workshop process, highlighting aspects of public understanding of residential photovoltaic technology.

12.1.4 PREVIOUS STUDIES OF MOTIVATIONAL AND ATTITUDINAL FACTORS ASSOCIATED WITH RESIDENTIAL GRID-CONNECTED PHOTOVOLTAIC TECHNOLOGY

Studies of understanding and attitudes to domestic installations of grid-connected photovoltaic systems have largely been conducted in Europe and America, and generally show that citizens have positive attitudes towards GCPV [20]. Some populations while having a positive attitude towards GCPV tended to confuse such installations with roof-mounted solar hot water [21]. However, only a small proportion of a population are prepared to invest in such technology. Oppenheim [22] refers to 1% of consumers in two American cities being willing to pay extra to have photovoltaic panels on their roof. Faier and Neame [7] note the lack of uptake of a combined photovoltaic and thermal solar panel system made available in England under a grant system that provided 50% of the installed capital cost.

12.1.5 BUILDING INTEGRATION OF PHOTOVOLTAICS

Building-integrated photovoltaic panels do not feature extensively in installations in Australia, whereas they do in the northern hemisphere. Social research on this aspect of photovoltaic technology is limited, with Sylvester [23] having studied simulation of photovoltaic filtering on windows in office buildings. Although the participants were in favour of the energy and associated energy cost savings of this technology, they were dissatisfied with the disruption to natural light in the building. Blewett-Silcock [24] studied public reactions to building-integrated photovoltaic technology in an English university. He examined whether participants found the materials attractive and found that this depended on the type of building they were asked to evaluate. Interestingly, the participants did not link the electricity they used in the office with that generated by the façade.

12.1.6 PERCEIVED ECONOMIC FEASIBILITY

Perez et al. [25] consider the perception of economic feasibility of photovoltaic electricity generation and solar hot water production. Initially, they considered short-term payback, the net cost divided by the first year energy cost savings, but quickly dismiss this as too simple and partial, in failing to take account long-term impacts of GCPV. Long-term return requires assessing the net present value (NPV) - the economic value over a product's lifetime, where they demonstrated a positive value. Importantly, in assessing NPV, they highlighted that current retail costs do not reflect some of the advantages of GCPV with dispersed production of electricity on site rather than undispersed electricity, the value of which they estimated as US$0.01 to US$0.06 more per kWh. Hence, comparisons with other technologies should compare the NPV of lifetime distribution costs. Similarly, Riedy [26] notes that, in general, transmission pricing regimes in Australia are biased against distributed generation such as GCPV while favouring undistributed generation such as coal fire production. Distributed generation sources such as GCPV should not be compared as though they use the transmission system.

12.1.7 FEED IN TARIFF

Wiginton et al. [27] argue that feed in tariffs have been the most effective government incentive program for encouraging domestic photovoltaic installation, and those countries which have introduced feed in tariffs have seen the greatest uptake. Mitchell et al. [28] maintained that the German feed in tariff scheme is more effective at increasing the share of renewables than the Renewables Obligation in England and Wales because it reduces risk more effectively for generators; such arguments are reiterated by Lesser and Su [29]. Schaefer et al. [30] reported favourable citizen attitudes towards feed in tariffs in a survey of views on domestic wind energy production in New Zealand and refer to similar findings in Canada and Japan. Maine and Chapman [31] demonstrated that during industry infancy, paying the spot price for electricity in South Australia provided insufficient incentive to take up solar electricity, suggesting that it was an unsuitable base on which to formulate a feed in tariff to promote PV installation.

12.1.8 REBATES AND SUBSIDIES

Haas et al. [32] have examined participants' motives in the Austrian roof-top program of the early 1990s and found that the rebate was an essential factor in the adoption of GCPV for 40% of participants. About 35% of participants would have purchased a system without a rebate. Haas et al. [32] regarded the high investment costs as a major barrier for broader market penetration. They noted a high willingness to pay for photovoltaics, which is above the level of cost-effectiveness but argued that this willingness is dependent upon rebates or other financial incentives. Haas [33] argued that rebates are an effective tool in expanding photovoltaic markets but that the rebates in almost all programs are too high and do not provide a sufficient incentive for a customer to find the most efficient system. Hence, while they may be appropriate for an early stage of market diffusion, rate-based incentives are the most effective tool for efficiently increasing GPCV use.

Chosen features for a social learning and attitude evaluation process for GCPV in NSW

The review of social science studies relevant to GCPV in NSW suggests that a relevant social learning and attitude evaluation process should

- discuss and demonstrate how domestic installations operate and assess participants' attitudes to GCPV and willingness to pay to have an installation in their residence;
- provide information and assess understanding of the economic and environmental impacts of distributed generation and evaluate participants' attitudes towards such impacts;
- evaluate participants' attitudes to the design and appearance of building-integrated panels;

- provide explanations on the operation of feed in rates in New South Wales, provide comparisons with other States in Australia and assess participants' attitudes towards and preferences for levels of feed in rates;
- provide information on the types of rebate that have been made available by the Federal government in Australia and assess participants' attitudes towards and preferences for rebates provided.

These content and evaluation areas were incorporated in the development of a deliberative workshop process.

12.2 METHODS

12.2.1 DELIBERATIVE WORKSHOP METHOD

Four deliberative workshops were conducted with citizens in the Illawarra region of New South Wales, Australia, implementing a social learning method for developing and assessing public understanding of residential photovoltaic installations. The participants were systematically selected [34] from the local telephone directory and paid Australian $50 for their participation in a 90-min workshop. The workshop was conducted with the aid of a PowerPoint audiovisual presentation, which included information about how photovoltaic panels operated and images of photovoltaic panels utilized in domestic installations. The participants were provided with a booklet which had a printed version of the slides contained in the audiovisual presentation.

The workshop commenced with a discussion of renewable energy and the various means by which it can be produced. This focused on the generation of electricity through wind turbines and a local system which utilized wave energy. The audiovisual presentation provided information on the basic operation of a photovoltaic panel. It also presented images of the basic installation of panels on a domestic rooftop with traditional mounting of panels on a rack.

The presentation then provided photographs of building-integrated panels which are more prevalent in Japan and Europe. The participants were asked to rate the attractiveness and design of four building-integrated types of panels and four non-integrated types of panels on 10-point semantic differential scales, where 1 was attractive and 10 unattractive; and on a second semantic differential, where 1 was not well designed and 10 was well designed. These different types of panels were represented by computer graphics which presented different types of generic panel structures. These included non-integrated panels on two upright short racks, panels on four long flat racks, flat panels on tiles; and integrated panels on short skylights, elongated skylights, panels which completely replace roof tiles and look like a flat iron roof, and panels which completely replace traditional roof tiles and look like roof tiles.

The presentation then focused on the financial aspects of installing panels on a domestic dwelling and the rebates which were available from the Federal government. In 2005, the Federal Government provided a rebate of Australian $7,920 for the installation of 1.5 kWh of panels (usually six panels), which typically cost Australian $14,000 to install, thus requiring an outlay of Australian $6,080. The state governments would typically pay for electricity generated, which was excess to domestic usage, at the same rate at which they sold electricity to residents, usually at Australian $0.125/kWh. The participants were told that this size of installation would typically generate 1,461 kWh of electricity per annum which would be worth Australian $4,566 over a 25-year period, the expected life of the panels. The scenario of a payment of Australian $.50/kWh was also discussed. This feed in rate was available in Germany and was being considered for the Northern Territory in Australia. They were also told that the maintenance cost of the panels would typically be Australian $650 for the purchase of a new inverter, which might become necessary at some stage of the expected 25-year life of the panels.

The participants in 2012 were told that a 1.5-kWh system would cost Australian $7,000 and there would be a government rebate in the form of a renewable energy certificate with a solar credit of Australian $3,720 which the supplier via the resident would receive, leaving a resident to pay Australian $3,280. They were told that if they used the electricity generated, depending on the tariff they paid for electricity from the grid, it would be expected to be worth Australian $300/year at Australian $0.20/kWh or Australian $7,500 over the expected 25-year operation of the panels, assuming electricity prices increase at the same rate as inflation. They were also told that this may not be feasible, as for it to be worth this amount, they would have to use energy during daylight hours, when it was generated. If they fed it back to the grid, then under proposed arrangements, they would receive about Australian $0.077/kWh, the wholesale undistributed price to be paid for electricity generated by PV in NSW. They were also told that the maintenance costs for a set of panels would typically be Australian $650 for the purchase of a new inverter at some stage of the expected 25-year working life of the panels.

The participants were finally asked to complete a questionnaire in which they rated their agreement with various attitude statements about photovoltaic panels, on 7-point Likert scales, where 1 indicated very strongly agreeing and 7 was very strongly disagreeing. They were asked to rate whether they considered the panels to be environmentally worthwhile, financially worthwhile, reliable, safe, whether or not panels could be constructed to be unnoticeable on roofs and whether or not they detract from the appearance of a house. They were also asked to rate their intention to install photovoltaic panels on a 4-point scale of very likely to very unlikely. If they were not prepared to install panels, they were asked if they would be prepared to do so if the rebate in 2005 or the solar credit in 2012 was higher and by how much, and if a feed in tariff rate on electricity generated was higher and by how much.

Discussion about the various technological and financial aspects of photovoltaic panels was encouraged in the workshop. Discussion was encouraged through asking the participants what they knew about renewable energy and how important it was to them; whether they had any safety concerns about installing a set of photovoltaic panels on their residence, what they thought of the appearance of integrated and non-integrated photovoltaic panels, what they thought of the financial benefits of installing a set of panels on their residence and whether they thought it was more important to receive a high rebate or feed in tariff.

Bang et al. [35] argue that more qualitative work is needed to ascertain what beliefs people have about renewable energy. The workshop proceedings were recorded, transcribed and qualitatively analysed, but while the quantitative results are reported, the qualitative results are not reported in detail in this particular article.

12.2.2 RESEARCH EXPECTATIONS

Although this was a preliminary study examining how a social learning method could be implemented on residential grid-connected photovoltaics, there were some expectations about the deliberations which ensued even though an initially small sample size was utilized. These expectations were as follows:

Only a small proportion of the public would be prepared to purchase a set of photovoltaic panels in 2005, and the awareness and understanding of photovoltaic technology would be minimal in the general community.

A higher proportion of the public would be prepared to purchase a set of photovoltaic panels in 2012, and there would be increased awareness in the local community about the nature of photovoltaic panels and familiarity with such panels partly due to the publicity surrounding the brief implementation of a subsidized gross feed in tariff policy, by the State government, which had been terminated before the 2012 workshops were conducted.

Positive attitude towards the environmental benefits of photovoltaic panels would be less of a predictor of intention to purchase a set of panels than positive attitude towards the financial benefits of purchasing a set of panels.

A larger proportion of the sample would be prepared to purchase a set of panels if the installed cost was lower and/or rebate in 2005 or solar credit in 2012 and feed in tariff rate was higher.

The public would have more positive attitudes towards the attractiveness and design considerations of building-integrated panels than traditional panels.

12.3 RESULTS

12.3.1 ANALYSIS OF QUESTIONNAIRE RESPONSES

In 2005, seven participants attended the first workshop, and eight participants attended the second workshop. In 2012, seven people attended the third workshop, and nine attended the fourth workshop. The participants comprised approximately half male and half female in each workshop, half of whom were over 45 years of age and predominantly living with a spouse or partner. The participants were from a range of occupational backgrounds with professional and retired categories, having more than five participants in 2005 and 2012. One third of the participants were receiving more than Australian $80,000 gross income per year, one fifth were receiving between Australian $50,000 and Australian $80,000/year, and one fifth Australian $40,000 to Australian $50,000/year. There were no differences in the demographic characteristics of the two samples in 2005 and 2012.

TABLE 1: Frequency of response of participants' likelihood of purchasing a photovoltaic system

Response	Year		Total
	2005	2012	
Very likely	2	1	3
Likely	2	7	9
Unlikely	7	7	14
Very unlikely	4	1	5
Total	15	16	31

None of the participants in the 2005 sample had purchased a photovoltaic system, whereas one participant had purchased such a system in the 2012 sample. While the mean difference in likelihood of purchasing a system between year of consultation (2005 M=2.86, SD=0.24; 2012 M=2.5, SD=0.22) was not statistically significant (F(1,30)=1.27, p<.28), in 2012, there was an 88% increase in the proportion of participants who said that they would be likely or very likely to purchase a photovoltaic system (8/16=50% vs 4/15=27%). There was a concomitant decrease of 77% in the proportion of participants (6% vs 27.%) who said that they would be very unlikely to purchase a photovoltaic system in the near future (see Table 1).

Forty percent (6/15) of the 2005 sample stated that they would need to receive Australian $0.50/kWh, and four participants chose other rates: Australian $0.20, Australian $0.30, Australian $0.70 and Australian $1/kWh, while 33% (5/15) of the participants in 2005 did not respond. In 2012, 38% participants responded (6/16) and were evenly spread over Australian $0.20 to Australian $1.00/kWh, while 62% (10/16) did not respond to this question. The mean difference between year of consultation in preferred feed in tariff rate was not statistically significant ($F(1,15)=0.57$, $p<.47$).

Participants' acceptable capital purchase price for a set of six panels are listed in Table 2. The mean difference in price that participants were prepared to pay for purchase and installation, between year of consultation, was not statistically significant ($F(1,20)=2.69$, $p<.11$). The median prices that participants were prepared to pay for the installation of a 1.5-kW system was significantly less in 2012 (Australian $2,000) compared to 2005 (Australian $5,000), consistent with a fall in capital cost for installing such systems. There was a large proportion of participants who did not respond to this question in 2012 (9/16), reflecting uncertainty around tariffs and return on investment from capital and the public framing of this decision in light of perceived losses in tariffs relative to earlier investment.

TABLE 2: Participants' preferred capital purchase price for a set of six panels

Purchase price (Australian $)	2005	2012	Total
0	0	1	1
500	0	1	1
1,500	0	1	1
2,000	2	2	4
2,500	2	1	3
2,800	0	1	1
3,000	1	0	1
4,500	2	0	2
5,000	2	0	2
6,000	4	0	4
30,000	1	0	1

Multivariate analysis of variance indicated a significant multivariate effect due to year of consultation on participants' mean ratings for the general characteristics of photovoltaic panels ($F(7,23)=556.85$, $p<.001$). The mean ratings of panels and the significance of the differences in mean ratings by analysis of variance (ANOVA) from 2005 and 2012 are shown in Table 3. The participants in 2012 placed significantly greater emphasis on whether investing in panels was financially worthwhile or not as well as statistically significant greater emphasis on whether panels detract from the appearance of houses.

TABLE 3: Participants' ratings of the characteristics of photovoltaic panels

Panel characteristic	2005 mean ratings (SD)	2012 mean ratings (SD)	F(1,29)	Probability
Environmentally worthwhile	1.87 (1.06)	2.12 (0.23)	0.59	<.45
Financially worthwhile	5.43 (0.19)	2.94 (0.17)	95.86	<.001
Not financially viable	3.29 (0.32)	4.53 (0.29)	8.38	<.01
Reliable	2.29 (0.27)	2.71 (.25)	1.29	<.27
Safe	2.00 (0.96)	2.65 (0.20)	4.69	<.04
Unnoticeable on roofs	2.00 (0.29)	3.35 (0.26)	12.03	<.01
Detract from the appearance of a house	4.71 (0.39)	3.59 (0.35)	4.63	p<.04

Ratings of whether or not a grid-connected photovoltaic installation was environmentally worthwhile and financially worthwhile were regressed on ratings of likelihood to purchase a set of panels, in order to assess the relative predictive efficacy of these factors. The regression model was significant ($F(2,30)=4.45$, $p<.05$). The likelihood of purchasing a set of panels was predicted by whether participants considered that panels are financially worthwhile and not based on whether they considered such panels to be environmentally worthwhile.

The mean ratings for the attractiveness and design of the integrated panels were that such panels were considered more attractive than non-integrated panels in 2005 and 2012, see Table 4. For the repeated measures ANOVA, there was no main effect for year of consultation or significant interaction between the rating of attractiveness and year of consultation. The integrated panels were also rated as being better designed than the non-integrated panels in 2005 and 2012. Once again, there was no main effect for year of consultation or significant interaction between rating of design and year of consultation.

TABLE 4: Participants' ratings of the attractiveness and design of building-integrated and non-integrated panels

Panel characteristic	2005 mean ratings (SD)	2012 mean ratings (SD)	$F_{(1,30)}$	P value
Attractiveness of integrated panels	3.66 (1.65)	4.19 (1.35)	35.60	<.001
Attractiveness of non-integrated panels	5.51 (1.93)	5.47 (1.61)		
Good design of integrated panels	8.28 (1.07)	6.74 (1.43)	34.97	<.001
Good design of non-i ntegrated panels	5.88 (1.96)	5.08 (1.25)		

Although the qualitative analysis of the workshop discussion is not reported here, it is worth noting that some 2005 workshop participants initially misunderstood photovoltaic technology to be solar hot water technology. The participants in the 2012 workshops did not indicate this misunderstanding but were concerned about the safety of installing photovoltaic panels on their roofs, whether such panels were a fire hazard and whether they would withstand storm damage. The qualitative data obtained in the workshops provide details of the social learning process which occurred. Briefly, the participants shared their knowledge of renewable energy, which was primarily solar hot water in 2005. Through discussing the difference between solar hot water production and photovoltaic electricity production, the participants shared and developed their understanding of the differences between these two forms of renewable energy. There was considerably more understanding of the operation of photovoltaic panels in 2012, and the participants were fairly conversant with the operation of such technology. Social learning also took place with regard to the building integration of photovoltaic panels. There was considerable discussion about how this type of photovoltaic panels was preferable to stand-alone panels in 2005 and 2012 in terms of the aesthetic benefits and in terms of the monetary benefits of substituting photovoltaic panels for building materials. The participants' understanding of the financial costs and benefits of GCPV also benefited from a social learning process, whereby participants shared their understanding of feed in tariffs and how they were operating in 2005 and 2012.

12.4 DISCUSSION

The questionnaire results indicate positive attitudes to domestic installation of photovoltaic panels but a reticence to participate in the rebate scheme in existence in 2005 because it was not seen as financially viable. The participants in 2005 were more willing to participate in the scheme if the government rebate had been higher. The participants also felt that the feed in tariff should be higher. The participants in the 2012 workshops were more likely to purchase a set of panels, but the difference with the 2005 participants was not statistically significant.

Forty percent of the participants in the 2005 workshop (6/15) indicated an interest in receiving Australian $0.50/kWh. This was probably due to a discussion in the 2005 workshops of an Australian $0.50/kWh hour feed in tariff rate proposed for the Northern Territory in Australia. The participants' preferred feed in tariff was evenly spread between Australian $0.20 and Australian $1.00/kWh for 2012, while 38% (6/16) did not respond. This indicates that the participants may not be connecting return on investment to initial investment capital and had difficulty answering the question. There was also a high non-response to how much they were prepared to pay for the purchase of a set of panels (32% in 2012). Non-responses were not observed for other items in the questionnaire. These questions may be more readily answered through a discrete choice experiment structure [36] which provides a more readily interpreted structure and such a structure will be trialled in future research. The high level of prices that participants were prepared to pay in 2005 (median of Australian $5,000) than those in 2012 (median of Australian $2000) is expected given the quoted retail price of Australian $6,080 when the federal government rebate was taken into account in 2005, whereas in 2012, the general price for a 1.5-kW system was Australian $3,280 when solar credits were taken into account.

The participants' ratings of the general characteristics of panels differed significantly between years of consultation. The participants in 2012 had greater agreement than the 2005 participants with the statement that panels are financially worthwhile and disagreed more with the statement that panels are not financially viable. This reflects the significantly lower capital costs of installing systems and despite 36% lower feed in tariff than in 2005 (Australian $0.08 vs Australian $0.13c/kWh). The participants in 2012 were concerned about the safety aspects of installing photovoltaic panels on a residence. This was reflected in the ratings that participants provided on the characteristics of panels and also in the discussions which took place in the workshops in 2012, where participants were concerned about whether or not the panels could cause a fire or whether or not they would withstand storm damage. This may have been instigated by the reports in the media in the year preceding the 2012 workshops about electrical and fire problems which had arisen through an environmental program in which the installation of house insulation was implemented through a subsidy provided by the Federal Government.

In 2005, the participants' understanding of solar hot water generation was the initial knowledge of parallel phenomena which they brought forth in their discussions about photovoltaic electricity generation. Some participants in 2005 did not initially understand the differences between solar hot water and photovoltaic electricity generation. Knowledge about solar hot water production was what they initially discussed in relation to the topic of the workshop.

Government action, encouraging public action, was also considered important by the participants in 2005 and 2012. In 2005, they considered that the subsidy should be greater to enable more panels to be installed. The participants in 2005 were also in favour of a higher feed in tariff. They considered that they would be more motivated to purchase panels if the feed in rate was higher. The participants in 2012 were still in favour of higher feed in tariffs as would be expected.

The semantic differential ratings of the building-integrated panels indicated a positive evaluation of such integration with participants, considering that such panels were more attractive and better designed in both 2005 and 2012. The ratings suggest that retailers should consider marketing and selling such panels as they may significantly enhance the public's interest in and attitude toward such technology.

Discussion in the 2005 workshops indicated misunderstandings in the participants' perceptions of the practicalities of utilizing GCPV technology. If the researchers had relied on their own understandings and knowledge of citizen interpretations of the way in which such technology operates, they would have missed out on various aspects of citizen understanding of photovoltaic technology which were tacit in their understanding of how such technology operates. This indicates the worth of implementing a social learning approach to informing the public about residential grid-connected photovoltaic installation.

Discussion in the 2012 workshops indicated greater understanding of the operation of domestic photovoltaic installations than that in the 2005 workshops. However, the research indicates that the 2012 participants had increasing difficulty expressing their attitudes towards the financial arrangements which are currently in place for residents who wish to install photovoltaic systems in their residences, which likely reflect changes in tariff policy and associated uncertainty. Importantly, such policy changes in NSW appear to have framed public perception of investment in GCPV now as a loss relative to investment when tariffs were Australian $0.60c/kWh until 2016. This is in spite of long-term returns on investment in GCPV in fact being higher as capital costs of installed GCPV systems have significantly fallen and despite NSW pricing not appropriately having taken into account the lower distribution costs of residential GCPV-generated electricity. This research is timely in informing the public and government policy, showing the impacts on decisions to install GCPV technology are in large part dependent on factors which have not been considered or researched. Namely, what citizens perceive as reasonable and fair and sufficiently motivating to invest in installing a GCPV system in their residences, how well that is informed and how those factors have been framed and changed over time [10,12].

12.5 CONCLUSIONS

Social learning principles can provide a range of benefits for communication and decision making in the informed promotion of grid-connected photovoltaic technology. Public perceptions and citizens' investment decisions should move beyond framing decisions relative to subsidy levels to consider long-term investment returns, which generally continue to improve with falling capital costs and higher energy prices, despite reducing generation subsidies in New South Wales since 2010. Retailers and installers of residential photovoltaic systems in Australia are encouraged to promote the option of building-integrated panels given favourable preferences shown for such technology.

REFERENCES

1. Bakos GC, Soursos M, Tsagsas NF (2003) Technoeconomic assessment of a building-integrated PV system for electrical energy saving in residential sector. Energ Build 35:757-762
2. Australian Government Clean Energy Regulator (2012a) RET power stations. http://ret.cleanenergyregulator.gov.au/For-Industry/RET-Power-Stations/ret-power-stations . Accessed 30 Nov 2013
3. Australian Government Clean Energy Regulator (2012b) Renewable energy target. http://www.cleanenergyregulator.gov.au/Renewable-Energy-Target/Pages/default.aspx . Accessed 1 Jul 2012
4. Macintosh A, Wilkinson D (2011) Searching for public benefits in solar subsidies: a case study on the Australian government's residential photovoltaic rebate program. Energ Policy 39:3199-3209
5. Wustenhagen R, Wolsink M, Burera M (2007) Social acceptance of renewable energy innovation: an introduction to the concept. Energ Policy 35:2683-2691
6. Wolsink M (2007) Planning of renewables schemes: deliberative and fair decision-making on landscape issues instead of reproachful accusations of non-cooperation. Energ Policy 35(5):2692-2704
7. Faiers A, Neame C (2006) Consumer attitudes towards domestic solar power systems. Energ Policy 36:1797-1806

8. Australian Federal Government (2011) Solar Cities: catalyst for change - Background paper. http://www.climatechange.gov.au/energy-efficiency/solar-cities . Accessed 30 Nov 2013

9. Australian Government Clean Energy Regulator (2012c) Renewable energy target: feed in tariffs. http://ret.cleanenergyregulator.gov.au/Solar-Panels/feed-in-tariffs . Accessed 1 Jul 2012

10. Frontier Economics (2012) Market value of solar PV exports: final report prepared for IPART. Melbourne: Frontier Economics Pty. Ltd..

11. Martin J (2012) NSW Solar Bonus Scheme benchmark rate range to be 7.7c-12.9c for 2012–2013: IPART. http://www.solarchoice.net.au/blog/nsw-solar-bonus-scheme-benchmark-rate-range-to-be-7-7c-12-9c-for-2012-2013-ipart/ . Accessed 1 Jul 2012

12. Independent Pricing and Regulatory Tribunal (2012) A fair and reasonable solar feed-in tariff for NSW. http://www.ipart.nsw.gov.au/Home/Industries/Electricity/Reviews/Retail_Pricing/Solar_feed-in_tariffs_-_2012-2013 . Accessed 30 Nov 2013

13. Kahneman D, Tversky A (1979) Prospect theory: an analysis of decision under risk. Econometrica 47(2):263-292

14. Fischer F (2000) Citizens, experts and the environment: the politics of local knowledge. Durham NC, USA: Duke University Press.

15. Schusler TM, Decker DJ, Pfeffer MJ (2003) Social learning for collaborative natural resource management. Soc Nat Resour 16(4):309-326

16. Keen M, Brown V, Dyball R (2005) Social learning: a new approach to environmental management. In: Keen M, Brown VA, Dyball R (eds) Social learning in environmental management, London: Earthscan. pp 3-21

17. Webler T, Kastenholz H, Renn O (1995) Public participation in impact assessment: a social learning perspective. Environ Impact Asses 15(5):443-463

18. Collins H, Evans R (2007) Rethinking expertise. Chicago: University of Chicago Press.

19. Collins H (2010) Tacit and explicit knowledge. Chicago: University of Chicago Press.

20. Genennig B, Hoffman V (1995) Sociological accompanying study of the 1,000 roofs - PV programme initiated by the Federal as well as country government in Germany. 13th European photovoltaic solar energy conference. Nice, France, 23–27 Oct 1995.

21. van Mierlo BC, Westra CA (1995) Attitude of occupants towards utility owned PV-sytems. 13th European photovoltaic solar energy conference. Nice, France, 23–27 Oct 1995.

22. Oppenheim J (1995) A program to demonstrate that consumers place value on environmentally benign electricity: residential rooftop PV. 13th European photovoltaic solar energy conference. Nice, France, 23–27 Oct 1995.

23. Sylvester KE (2000) An analysis of the benefits of photovoltaic-coated glazing on owning and operating costs of high rise commercial buildings. Dissertation Abstracts International Section A. Humanities and Social Sciences 2(1-A):

24. Blewett-Silcock T (2000) Public reaction to building integrated photovoltaics. 16th European photovoltaic solar energy conference. Glasgow, UK, 1–5 May 2000.

25. Perez R, Burtis LA, Hoff T, Swanson S, Herig C (2004) Quantifying residential PV economics in the US payback vs. cash flow determination of fair energy value. Sol Energy 77:363-366

26. Riedy C (2003) Subsidies that encourage fossil fuel use in Australia. Institute for Sustainable Futures: University of Technology, Sydney.
27. Wiginton LK, Nguyen HT, Pearce JM (2010) Quantifying rooftop solar photovoltaic potential for regional renewable energy policy. Comput Environ Urban 34(4):345-357
28. Mitchell C, Bauknecht D, Connor PM (2006) Effectiveness through risk reduction: a comparison of the renewable obligation in England and Wales and the feed-in system in Germany. Energ Policy 34:297-305
29. Lesser JA, Su X (2008) Design of an economically efficient feed-in tariff structure for renewable energy development. Energ Policy 36(3):981-990
30. Schaefer MS, Lloyd B, Stephenson JR (2012) The suitability of a feed-in tariff for wind energy in New Zealand—a study based on stakeholders' perspectives. Energ Policy 43:80-91
31. Maine T, Chapman P (2007) The value of solar: prices and output from distributed photovoltaic generation in South Australia. Energ Policy 35(1):461-466
32. Haas R, Ornetzeder M, Hametner K, Wroblewski A, Hübner M (1999) Socio-economic aspects of the Austrian 200 kWp-photovoltaic-rooftop programme. Sol Energy 66(3):183-191
33. Haas R (2002) An international evaluation of dissemination strategies for small grid-connected PV systems. World Renewable Energy Congress VII, Cologne, Germany, 29 June–5 July 2002.
34. Henry GT (1990) Practical sampling. In: Applied social research methods series 21: Newbury Park: Sage,
35. Bang HK, Ellinger AE, Hadjimarcou J, Traichal PA (2000) Consumer concern, knowledge, belief and attitude toward renewable energy: an application of the reasoned action theory. Psychol Mark 17(6):449-468
36. Hoyos D (2010) The state of the art of environmental valuation with discrete choice experiments. Ecol Econ 69:1595-1603

AUTHOR NOTES

CHAPTER 1

Acknowledgments

The research work described in this paper was fully supported by the France/Hong Kong Joint Research Scheme (project no. F_HK05/11T).

CHAPTER 2

Acknowledgments

We thank The Nature Conservancy for supporting this analysis. W. Christian, J. Kiesecker, M. Kramer, B. McKenney, J. Moore, E. O'Donoghue, S. Parker, C. Pienkos, J. Randall, R. Shaw, M. Sweeney, J. Ziegler, and especially L. Crane provided helpful discussions and review of earlier versions. Two anonymous reviewers also provided insightful and helpful comments which greatly improved the paper. We also appreciate the help of Catherine Darst in supporting our use of desert tortoise model data and providing Desert Tortoise Conservation Area data.

Author Contributions

Conceived and designed the experiments: DC SM. Performed the experiments: DC BC. Analyzed the data: DC BC. Wrote the paper: DC SM.

CHAPTER 3

Acknowledgments

We thank Konstantin Daskalov, Hiral Patel, Ashkan Foroughi, Daniel Frochtzwajg, Karo Karapetyan, and Michael Darling for experimental assistance. This work is dedicated to the memories of Michael Dickson and Andrew Wieting.

Author Contributions

Conceived and designed the experiments: CK YP HP. Performed the experiments: CK YP HP. Analyzed the data: CK YP HP. Contributed reagents/materials/analysis tools: CK YP HP. Wrote the paper: CK YP HP.

CHAPTER 7

This paper is a product of the Environment and Energy Team, Development Research Group. It is part of a larger effort by the World Bank to provide open access to its research and make a contribution to development policy discussions around the world. Policy Research Working Papers are also posted on the Web at http://econ.worldbank.org. The author may be contacted at gtimilsina@worldbank.org.

We sincerely thank Manu V. Mathai, Ashok Kumar, Jung-Min Yu, Xilin Zhang, Jun Tian, Wilson Rickerson, and Ashish Shrestha for research assistant and Ionnis Kessides, Mike Toman, Chandrasekar Govindarajalu, Mudit Narain and Katherine Steel for their comments. We acknowledge the Knowledge for Change Program (KCP) Trust Fund for the financial support. The views expressed in this paper are those of the authors and do not necessarily represent the World Bank and its affiliated organizations.

CHAPTER 9

Acknowledgments

This project was funded by the U.S. Department of Energy Solar Energy Technology program. Model datasets in this analysis were derived from publicly available sources; individual utilities modeled in this work did not provide proprietary data or review our interpretation of the data, model, methods, or results. The authors would like to thank Victor Diakov for performing the price sensitivity runs and Craig Turchi for performing the SAM CSP runs. The following individuals provided valuable input and comments during the analysis and publication process: Nate Blair, Adam Green, Udi Helman, Trieu Mai, Mark Mehos, and Frank Wilkins. Any errors or omissions are solely the responsibility of the authors.

CHAPTER 10

Competing Interests

The authors declare that they have no competing interests.

Author Contributions

AH contributed with the science explanation and background of the photovoltaic technology. AJA participated in the design of the model city, performed economic analysis of fossil-based fuel that was used as a benchmark, and helped to draft the manuscript. KJP performed the economic analysis of household photovoltaic technology. JPO provided the background explanation on how Dish Stirling Systems work and contributed with the economic analysis of concentrated solar power technology. MF participated in the design of the model city and the economic analysis of fossil-based fuel. VAB compiled the data of fixed and variable costs for implementing the solar farm, contributed with the economic analysis of concentrated solar power technology, and helped draft the manuscript. All authors participated in the design of the study, performed the final analysis, read and approved the final manuscript.

Acknowledgements

Professor Stephen Berry for revising this paper and guiding us every step of the way. Sean Gallagher for providing us with specific data on the fixed and variable costs of dish Stirling engines, its implementation, and its maintenance. Jim Christensen for providing us with detailed information on electricity consumption, substation calculations and insights into creating our model city

This article has been published as part of Chemistry Central Journal Volume 6 Supplement 2, 2012: Roles for chemistry in the world's energy problems. The full contents of the supplement are available online at http://journal.chemistrycentral.com/supplements/6/S1.

CHAPTER 11

The views expressed in this report are those of the authors, and do not represent the views or positions of the U.S. Department of Energy, the National Renewable Energy Laboratory, the Renewable and Sustainable Energy Institute, nor the University of Colorado.

Notice

This report was prepared as an account of work sponsored by an agency of the United States government. Neither the United States government nor any agency thereof, nor any of their employees, makes any warranty, express or implied, or assumes any legal liability or responsibility for the accuracy, completeness, or usefulness of any information, apparatus, product, or process disclosed, or represents that its use would not infringe privately owned rights. Reference herein to any specific commercial product, process, or service by trade name, trademark, manufacturer, or otherwise does not necessarily constitute or imply its endorsement, recommendation, or favoring by the United States government or any agency thereof. The views and opinions of authors expressed herein do not necessarily state or reflect those of the United States government or any agency thereof.

CHAPTER 12

Competing Interests

The authors declare that they have no competing interests.

Author Contributions

GH designed the study and questionnaire, and conducted the statistical analysis. SE provided analysis and commentary on the economic and financial aspects of the study. GH conducted the workshops in 2005 and 2012, and SE contributed to the 2012 workshops. The authors read and approved the final manuscript.

INDEX